P.O 1148524   C.2   181   19.2.82

C0-APN-102

Western Office,
Mining Research Laboratories,
Canmet,
3303 - 33 St. N. W.,
Calgary, Alberta.
T2L 2A7

TECHNOLOGY OF HYDROMINING AND HYDROTRANSPORT OF COAL

by

N.E. Ofengenden

and

A.G. Dzhvarsheishvili

TRANSLATION FROM RUSSIAN BY

Albert L. Peabody
Annapolis, Maryland

TRANSLATION EDITED BY
William C. Cooley
Rockville, Maryland

PUBLISHED BY

TERRASPACE INC.

304 North Stonestreet Avenue
Rockville, Maryland 20850

The first English edition of the Russian work:

TEKHNOLOGIYA GIDRODOBYCHI I
GIDROTRANSPORTIROVANIYA UGLYA
(Nedra Press, Moscow, 1980)

Copyright © 1981 Terraspace Inc.

This book or parts thereof, may not be
reproduced in any form without the written
permission of the publisher. All rights reserved.
ISBN 0-918990-08-4
Library of Congress Catalog Card Number 81-52864

TABLE OF CONTENTS

i

THE TECHNOLOGY OF HYDRAULIC MINING AND HYDROTRANSPORT OF COAL

Abstract

This book studies the status and prospects for the development of hydraulic mining and hydrotransport. A technical and economic foundation is provided for the areas of expedient application of hydrotransport, particularly for the delivery of coal to large users and the creation of fuel and power systems; hydrotransport equipment and conditions of its reliable operation are described.

The book is intended for engineering and technical workers involved in research, planning and operation of hydrotransport installations.

Tables 22, Figures 50, References 27.

Foreword

One progressive technological trend in the coal industry which is planned for introduction in the next few years is hydraulic mining and hydraulic transportation of coal, which in many cases can successfully compete with traditional technological processes. The long term operation of hydraulic mines and hydrotransport installations for the transportation of coal in the USSR and abroad has indicated that hydraulic technology, based on hydrotransport, is a progressive technological trend.

Hydrotransport using hydraulic technology is favorably distinguished by the fact that it allows various systems of working of coal deposits to be used, creating continuous flow technological systems.

Such specifics of hydrotransport as the possibility of full mechanization and automation, high productivity, low cost, significant reductions in the injury rate in the transportation of coal, allow good technical and economic indices to be achieved.

When hydrotransport is used in hydraulic mines, it is possible to use systems of working with extraction of coal without the constant presence of personnel at the extraction faces in hydraulic mines.

The reliability of hydrotransport and hydraulic hoisting processes, a continuous and inexpensive form of transport with high productivity requiring no load transfer operations, is greater than that of other transportation equipment used in mines. Hydraulic mines have practically zero injury rates and avoid diseases related to dust formation. These advantages of hydromechanization are so great that they alone could require broader utilization of hydraulic mining technology. Furthermore, hydro-mechanization greatly simplifies the surface of a mine.

As a result of studies in the laboratory and particularly under field conditions, experience has been gained in the operation of hydrotransport allowing effective technological systems and equipment yielding the greatest technical and economic effect to be determined. In the Kuznets and Donets Basins, technological systems and equipment for hydrotransport, mining and processing of coal with hydraulic technology has been utilized on a broad industrial scale. In hydraulic mines with transportation distances of up to 3-4 km, gravity flow transport is used, with greater distances, pressurized transport is used. On the surface at mines, pressurized transport is generally used. Coal is extracted by mass-produced hydro-monitors and mechanical-hydraulic combines with remote control. Beneficiation, dewatering of coal at hydraulic mines, as well as the clarification of return water, are performed by equipment similar to the equipment used in plants for wet process coal beneficiation.

A technology of hydraulic and mechanical-hydraulic extraction of coal has been implemented; explosive plus hydraulic extraction, which represented over 50% of all hydraulic mining ten years ago, is not in use in any of the hydraulic mines in the nation today.

Four new types of monitors, six types of coal pumps, fittings and valves for slurry lines and water lines in hydraulic mines are currently in series production. The reliability of hydraulic equipment has been improved: that of water and coal pumps by a factor of 2, valves by a factor of 3, monitors and centrifuges by a factor of 5. The new equipment has allowed the reliability and effectiveness of hydrotransport at hydraulic mines to be improved, and has also resulted in improvement in the technology of driving of development and other workings.

Large hydraulic mines with beneficiation plants have been put in operation: Krasnoarmeyskaya, and 50th anniversary of the USSR (Donets Basin), and Yubileynaya and Inskaya (Kuznets Basin).

Over the past decade in the Kuznets Basin, a hydrotransport line 10-12 km has been successfully put into operation, a portion of a fuel and power system (Inskaya Hydraulic Mine, Belovskaya Regional Heat and Electric Power Plant) plus a fuel and metallurgical system (Yubileynaya Hydraulic Mine - Western Siberian Metallurgical Plant).

The experience which has been gained indicates that fuel-power systems consisting of a hydraulic mine, hydrotransport system and thermal electric power plant are quite promising.

Large hydraulic mines with high productivity, completely mechanized and automated, can be constructed as a component part of a fuel and power system in which the fuel supplier and consumer are connected by hydrotransport.

During the past 10 years the hydraulic technology of mining and transportation of coal has been increasingly used, its volume having grown by a factor of 2.5.

During this period, the hydraulic method in this nation has produced 60 million tons of coal, approximately 25 million tons in the Donets and 35 million tons in the Kuznets Basins.

The mines of the USSR in 1977 included 10 hydraulic mines, 5 in the Donets Basin and 5 in the Kuznets Basin (Table 1).

We present below some data on the dynamics of the growth of hydraulic coal mining in the USSR [1, 2].

Years	1952	1954	1958	1962	1966
Production, thousands of tons	15.1	265.9	1026.4	2814.2	4131.9
Year	1970	1972	1975	1979	
Production, thousands of tons	9097.9	10095.7	9311.7	8900	

Further development of hydrotechnology requires summarization of experience in order that planning organizations may obtain the required initial data for the selection of hydraulic mine, hydrotransport and fuel-power system projects.

This book attempts to summarize the experience gained in the operation of the best hydraulic mines and to note trends in the development of hydrotechnology for the creation of long distance hydrotransport systems for the transportation of coal from mines to large consumers in order to achieve broader introduction of new and progressive technology.

In this book Chapters 1-6 and 9-10 were written by Professor N. Ye. Ofengenden, Chapters 7-8 by Professor A. G. Dzhvarsheishvili.

Table 1.1

Hydraulic mine, production union (pu), region (rn), mine administration (ma)	Annual capacity, $t \cdot 10^3$	Pitch of seam, degrees	Average thickness of seam, m	Total output of coal in 1976, $t \cdot 10^3$
Kuznets Basin				
Krasnogorskaya, pu Gidrougol', Prokol'evsko-Kiselevskiy rn	650	45-70	2.63	651.6
Zarechnaya, pu Gidrougol', Leninsko-Belovskiy rn	750	5-6	2.17	862.7
No. 2 Inskaya, pu Gidrougol', Leninsko-Belovoskiy rn	850	19-35	2.21	2113.3
Yubileynaya, pu Gidrougol', Baydayevskiy rn (Yuzhnyy Kuznets)	3250	0-38	2.28	3580.5
Tyrganskaya, pu Gidrougol', Prokol'evsko-Kiselevskiy rn	300	50-80	10.00	1416.9
Donets Basin				
Krasnoarmeyskaya, pu Donetskugol', Krasnoarmeyskiy rn	1350	10-12	1.22	1563.5
Im. 50-letiya SSSR, pu Krasnodonugol', Krasnodonskiy rn	700	0-45	1.30	934.8
Pioner, pu Donetskugol', Krasnoarmeyskiy rn	780	10-13	0.84	733.7
No. 4 Ordzhonikidzeugol', Aleksandrovskoye ma	70	60-75	1.6	245.1
No. 105, pu Krasnoarmeyskugol', Kurakhovskoye ma	190	9-12	1.29	323.6

CHAPTER 1. TECHNOLOGY OF HYDRAULIC MINING OF COAL

1.1. Technological Systems of Hydraulic Mines

One necessary element of the process performed at all hydraulic mines is hydrotransport of the coal within the mine. The following are the most important trends in the application of hydromechanization which we have seen to date.

The first trend is hydraulic breaking of coal with monitors using jets of water at pressures of up to 10-16 MPa with average flow rates of water of 300 to 500 m³/hr. The broken coal is transported by gravity to a shaft and then by various means (coal pumps, feeders or air lifts) is raised to the surface. This trend can be considered among the most promising, since the coal is delivered from the mine to the consumer in a small number of steps.

A second trend is represented by breaking of the coal with hydro-mechanical machines. Hydrotransport and hydraulic hoisting of the coal involve the same processes as are used with hydraulic breaking.

A third trend is the use of hydraulic transport with mechanical breaking of the coal by standard means using the same technology of hydrotransport and hydraulic hoisting.

Breaking of the coal with water or other means, followed by feeding of the slurry through flumes in the openings to a central hydraulic hoist is the most commonly used technological system for hydraulic mining. Large lumps of coal must pass through crushers before they enter the coal pump. The slurry is fed to the beneficiation plant through pipes, then is dewatered (on screens, in centrifuges, etc.). The water is clarified and sent to reservoirs, from which it is returned to the mine.

The technological systems of hydraulic mines may involve simultaneous output of coal without separating the larger lumps or separate output of coal of different particle sizes.

Within a mine field, gravity hydrotransport is used to carry the coal to the shaft, regardless of the technological systems used. In technological systems with combined output, the coal is carried to the surface and transported to the beneficiation plant or consumer as a slurry.

In systems involving separate transport, the coal is separated on vibrating screens at the shaft. Coal particles more than 6 mm in diameter are carried to the surface in rail cars, skip cars or conveyors, while the slurry of the finer particles is thickened in cyclones and hoisted to the surface by coal pumps or feeders. The water clarified in the cyclones is returned by pumps to the mining faces for reuse in hydrotransport.

The simplest technological system involves a single operating system from mine face to consumer. The coal broken by the jet from the monitor is carried by gravity flow to the shaft and hoisted to the surface by various devices (coal pumps, air lifts, feeders), then goes on to the consumer. All operations in the cycle of breaking and transportation of the coal are performed by water.

Hydraulic mining systems have been used in which the coal is broken by hydromechanical equipment, transported within the mine by gravity hydrotransport, then hoisted to the surface by the same methods used in systems with hydraulic breaking.

The technological systems for hydrotransport of coal in hydraulic mines consist of the following elements: hydrotransport at the face and through the main openings to a central hydraulic hoist; the equipment which prepares the coal for hydrotransport; the hydraulic hoist which raises the coal to the surface; the hydrotransport water supply system; and the system which receives the coal at the surface.

Depending upon a number of factors (system used to expose seams, output of hydraulic mine, properties and eventual use of coal, specifics of surrounding rock, etc.), each of these elements may vary in extent.

Figure 1.1 shows a simple technological system with hydraulic breaking of the coal, hydrotransport under pressure through main openings and dewatering of the coal on a screen (without a beneficiation plant). The coal broken by the monitor jet is mixed with water and flows by

2

gravity as a slurry through flumes laid on a gravity incline to a sump at a coal pumping station for the section. As a rule, gravity hydrotransport is used in hydraulic mines at the face as the simplest and most reliable type. The slurry is fed through roadways by pressurized hydrotransport using coal pumps to the central hydraulic hoist. Pressurized hydrotransport is used to carry the slurry through primary openings which have the same slope as those used in ordinary mines.

Before the slurry is fed to the coal pump it passes through a crusher in which the maximum particle size is reduced to 50-60 mm. When this system is used, no additional crushing of the coal is performed before it enters the flume.

The coal is hydraulically hoisted to the surface using coal pumps located in a central chamber near the shaft.

On the surface, the slurry pours out onto a screen; the top product is sent to a coal washing combine, the bottom product to a reservoir. After the slurry has settled, the water is pumped back to the mine to the monitors, while the coal from the reservoir is fed to drainage areas.

The coal is dewatered at the drainage areas. In the summer this method of drying the coal is effective, but in the fall and winter it causes great difficulties.

The moisture content of the slurry dumped onto the drainage area should not exceed 35%; otherwise the slurry will flow off of the area. Coal slimes are dried to 20% moisture content on the drainage areas.

This type of hydraulic mine system has comparatively low output and has achieved only limited use.

A more productive hydraulic mine technological system is shown in figure 1.2. The coal is mined with powerful monitors and coal mining machines. The slurry is sent through flumes to crusher-classifiers.

VNIIGidrougol' Institute has developed the DKU crusher-classifier, which consists of a nonmoving grizzly, screen and hammer crusher. Materials which cannot be crushed are removed from the crusher.

Industrial operation has demonstrated that these machines are reliable, simple, and require cleaning of the vertical portion of the screen to

3

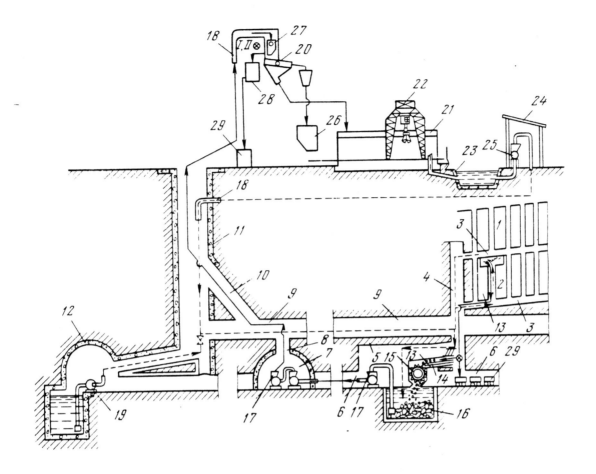

Figure 1.1. Simple hydraulic mine technological system: 1, 2--mining faces; 3--cross; 4--gravity incline; 5--section coal pump; 6--transportation roadway; 7--central coal pumping plant, 10--crosses; 9--ventilation opening; 11--supplementary shaft; 12--water drainage chamber; 13, 14--shutes; 15--crusher; 16--sump; 17--coal pump; 18--pipe; 19--pump; 20--screen; 21--settling tank; 22--gantry crane; 23--reservoir for clarified water; 24--feeder station; 25--feeder pump; 26--coal washing machine; 27-29--sampling devices.

remove foreign objects no more than twice per shift. The installation can crush rock in the 0-80 and 0-25 mm particle size range. At present, rock mass is being crushed at "Inskaya" and "Yubileynaya" hydraulic mines to 0-25 mm diameter. The slurry is sent to the sumps of the central hoist and transported by coal pumps to the beneficiation plant, where it is processed, while the clarified water is sent to the water supply and purification section and transported by pumps back to the monitors in the mine.

This system is distinguished by the fact that it uses gravity hydro-transport from the face to the central hydraulic hoist. This type of transport has been found more reliable and has withstood long term testing in industry. The slurry preparation unit is shifted to the central hydraulic hoist, allowing it to operate more effectively. The slurry is delivered to the beneficiation plant where it is treated by more sophisticated methods than those used in the other systems we have described. However, the DKU system does have its shortcomings, which reduce its operational reliability. Large metal and wooden objects must be removed from the intake aperture of the machine by hand; furthermore, the intake aperture is quite high and therefore it cannot be used at many operating hydraulic mines.

VNIIGidrougol' has developed a technological system for a modern hydraulic mine using series produced equipment. Coal is broken at the mining faces using the K-56 mg mining machine or one of its modifications.

Coal may also be broken using GMDTs-3M monitors with remote control. These monitors are suitable for mining of seams at least 0.8 m thick with a dip angle of over 6°. The monitor has an output of over 50 tons per hour at an operating pressure 12 MPa, water throughput up to 150 m^3/hr for coal with a (Protod'yakonov) hardness of 0.8-1. Type 12GD-2 monitors are used to mine coal in moderately thick and thick seams by remote control. Their production rate for coal with a hardness of 0.8-1 is 70 tons per hour assuming the water pressure is 12 MPa and discharges 400 m^3/hr.

The simplest and most reliable method of mining coal, when the conditions are right, is hydraulic breaking using jets with high flow

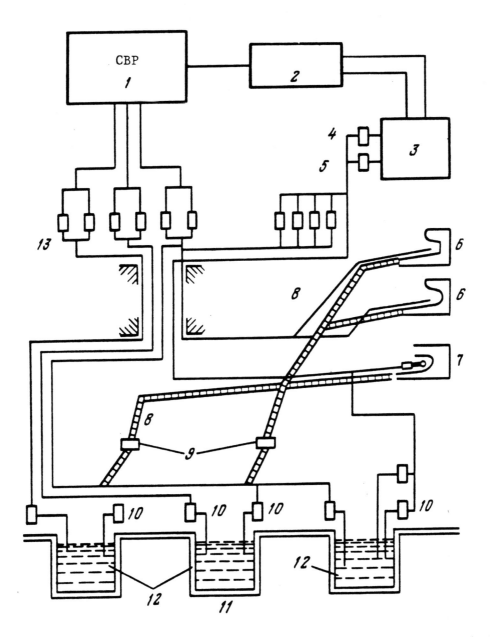

Figure 1.2. Technological system of a modern highly productive hydraulic mine: 1--central beneficiation plant; 2--water supply and purification section for process water; 3--process water distributor; 4--ZGM-2M soil pumps; 5--12 MSG-7 x 8 high-head pumps reinforced with hard alloy (delivery 800 m^3/hr, H = 100 MPa); 6--12 GD monitor (Q = 300 m^3/hr, H = 10-12 MPa, P = 100-140 t/hr); 7--K-56 MG mining machine (P = 80 t/hr); 8--flume for nonpressurized hydrotransport; 9--DKU crusher-classifier; 10--14 UV6 coal pumps for hydraulic hoisting and feeding of coal to beneficiation plant; 11--hydraulic hoisting plant; 12--sump of hydraulic hoisting installation; 13--transfer station to feed coal slurry to beneficiation plant.

rates. Strong and tough coal is usually broken using hydromechanical machines.

The selection of a system of mining must consider the tendency of the surrounding rock to absorb the water. If there is little tendency toward water absorption, hydraulic breaking with mobile remote control monitors is quite suitable. Technological systems for working of seams of coal with a tendency toward spontaneous combustion must be evaluated as to speed of mining and operating losses. At present, the losses of coal may be as high as 20-30%. A significant fraction of these losses result from failure to wash all the coal down; the losses can be reduced by the use of hydromechanical machines, which significantly reduce the losses of coal both through the thickness of the seam and over its area.

The hydraulic method of mining of coal, first used in the Soviet Union on the suggestion of Professor V.S. Muchnik, has now come into use abroad. For example, in England in 1969 experiments were conducted on the introduction of the hydraulic method of coal mining. The hydraulic method was used to drive openings, between which mining crosses 2 to 4.5 m in width were cut at a distance of 6.5 m from each other. Technological systems developed in the Soviet Union were used. The water pressure at the monitor nozzle was 7.3 to 8.7 MPa, flow rate 117 m^3/hr. The maximum production rate was 41.5 t/hr, maximum width of each pass 4.2 m. Each monitor produced up to 72 tons per shift, the total volume of coal mined per shift reaching 287 t. Hydraulic mining was found to be economically effective under these conditions.

1.2. Technological Equipment and Methods of Extraction of Coal

Hydraulic mines primarily use hydraulic and hydromechanical mining of coal. Effective hydraulic breaking of most seams in shaft mines can be achieved with a pressure of 12-16 MPa.

Higher pressures in many cases may be economically unjustified if they must be created by pumps on the surface. It is more desirable to use a pressure booster located near the mining face.

When the hydraulic method of mining is used, the cost of electric power amounts to 60-70%. It has been found that some 65% of all of the pressurized water expended is used in the operation of undercutting the face.

The output of a monitor cannot be theoretically calculated, since it depends on a large number of factors which cannot be strictly considered in theoretical computations. They include the physical and mechanical properties of the rock, the specific features of the mining system and many others. Experiments which have been performed have demonstrated that there are factors which can be considered. For example, it has been established that the output of a monitor depends on the kinetic energy of the jet and the distance from the monitor to the face.

The output of hydraulic breaking, ignoring mine factors, can be determined by a method developed by Doctor of Technical Sciences N.F. Tsyapko [3]

$$\theta_h = \frac{Q_p}{36.7 E_t} = \frac{1.02 \cdot 10^3 \, d^2 \, p\sqrt{p}}{E_t} \quad \text{t/hr} \tag{1.1}$$

where Q is the flow rate of water through the nozzle, m^3/hr; p is the pressure of the water at the nozzle, MPa; d is the nozzle diameter, m; E_t is the total power consumed in hydraulic breaking.

$$E_t = 0.2 \, E_1 + 0.8 \, E_3, \quad \text{kW} \cdot \text{hr/t}$$

$$E_1 = 4.3 \cdot 10^3 \, \frac{f_t^2 \, \sqrt{f_t}}{p} \quad \text{kW} \cdot \text{hr/t}$$

$$E_3 = 24 \, \frac{f_t^2}{\sqrt{p}} \quad \text{kW} \cdot \text{hr/t}$$

where $f_t = f\psi$ is the adjusted coal hardness factor; f is the Protod'yakonov hardness of the coal; ψ is the jointing factor of the coal (for the conditions of the Kuznets basin $\psi = 0.7-1$).

For cutting and development faces in hydraulic mines of the Kuznets basin, the output of hydraulic breaking (θt) considering the average

8

time spent in washdown, removal of oversize lumps and inspection of the face,

$$\theta_t = 0.85 \; \theta_h, \; t/hr \tag{1.2}$$

The specific energy consumption of hydraulic breaking is determined primarily by the mechanical strength of the coal, and for typical strengths without squeezing of the coal, varies from 0.85 to 14.4 kW · hr/m^3

Hydraulic breaking of coal depends not only on the degree of metamorphism, but also on the natural jointing. For coal with a high degree of metamorphism (type "A") hydraulic breaking is quite difficult. Coal in a moderate stage of metamorphism (types "K" and "PZh") is more brittle and soft and therefore is easily broken by jets of water at pressures of 4-6 MPa. Coal with a lower degree of metamorphism, with high strength and toughness (types "G" and "D") is not suitable for hydraulic breaking.

The effectiveness of breaking is greatly influenced by jointing. It has been found that the best economic results are achieved for hydraulic breaking of extraction faces, with poorer results at development faces.

The output of a monitor at extraction faces is 1.5 to 2.5 times higher than in development. This is explained by the influence of rock pressure, the larger exposed surface and greater width covered.

There is an optimal pressure at which the power consumption of breaking is minimal. As the diameter of the nozzle increases, the depth of the kerf and width of the chip both increase. There is thus an optimal value of nozzle diameter, achieving the maximum effectiveness of breaking. There are also minimum distances from nozzle to face, assuring the minimum specific flow rate of water for hydraulic breaking.

The effectiveness of hydraulic breaking is higher the more compact the jet and the higher the dynamic pressure in the core of the jet. The peripheral portion of the jet, where the pressure is less than it is on the axis, plays almost no part in breaking. The larger the diameter of the nozzle, the greater the portion of the jet which is not effectively used. The minimum specific flow rate of water can be achieved by increasing the dynamic axial pressure of the jet, which requires that the jet be compact, since its peripheral portion does not participate effectively in the breaking process.

Jet-shaping mechanisms called monitors with remote, semi-automatic or manual control, are used to break coal. Monitors with remote or semi-automatic control are designed to break the coal without requiring humans to be present at the face.

Monitors with nozzle diameters of 17-32 mm at pressures of 10-12 MPa and water flow rates of 150-390 m^3/hr are presently in use. Jets with pressures of not over 12 MPa can be used for more than 60% of the seams in shaft mines, while an increase in pressure to 16 MPa expands their area of application to 85-90% of all seams. In recent years a trend has been noted toward decreasing nozzle diameters to 6-9 mm and increasing pressures to 30-50 MPa.

A remote controlled installation based on a GMDTs-3 monitor has been widely used in the mines. It develops a pressure of up to 12 MPa and has wear-resistant sealing units. The GMDTs-3M monitor, operating in combination with the K-56 MG mining machine, has achieved excellent results in terms of output and extraction.

Further improvement of the theory of jet breaking of coal has allowed the creation of the GMDTs-4 mobile remote controlled monitor, with pressure up to 16 MPa and flow rate up to 180 m^3/hr.

In 1974, VNIIGidrougol' tested the 12 GD remote controlled monitor, designed for mining of coal in seams over 1.2 m thick with dip angles of over 6 degrees. The 12 GD monitor has a larger cross section of the channels through which the water flows, but otherwise is similar to the GMDTs-3M monitor.

Production by the 12 GD monitor is two to three times greater than that of the GMDTs-3M series produced monitor, as is the throughput of water. The length of the jet is increased with increased nozzle diameter, allowing operation with sublevel heights of up to 17 m. Work is now proceeding on decreasing the weight of this monitor.

Use of the 12 GD monitor with increased water flow rate has allowed an increase in the width and height of extraction columns (sublevels) by a factor of 1.5 to 2, achieving outputs of 100-150 t/hr at the face in weak and moderately strong coal.

Figure 1.3 shows the hydraulic characteristics of the 12 GD and GMDTs-3M monitors, obtained in bench and field tests.

The design and manufacture of self-propelled monitors with remote control and the 12 GD-2 monitor on the K-56 mining machine chassis has allowed effective operation in large openings.

Breaking of coal of practically any strength has become possible by the use of piston type pressure boosters, which develop jet pressures of up to 36-40 MPa.

High pressure hydraulic breaking of coal using GMDTs-3M monitors with flow rates of up to 180 m^3/hr of process water has continued the mean annual rates of increase in production at the face of about 5%. The use of the 12 GD-2 monitor, with its flow rate of up to 390 m^3/hr has allowed the output from the face to be increased to 68 tons.or 19%. It was also found possible to increase the height of a sublevel to 12-15 m (instead of 7-8 m with the GMDTs-3M monitor). The width of each pass is usually taken as 5-6 m, allowing the hourly output to be increased in proportion to the flow rate of process water, increasing the output of the face by a factor of 1.5 to 2. The pressure of the jet can increase to 16 MPa or higher. Thus, increasing the pressure from 10 to 13.5 MPa increases the output of the face by a factor of 3.5 to 4.

The 12 GP-1 monitor with track drive, developed by VNIIGidrougl', is currently under testing. This device allows mining of coal and cutting of openings. The monitor is moved by means of a powerful water turbine or electric motor. The operation of the monitor can be programmed. Upon completion of a mining pass, the monitor independently backs out of the face.

The GPS-1 monitor is program controlled for operation in both development and mining openings in seams of moderate thickness as well as thick seams with dip angle up to 15°. The operating water pressure is up to 10 MPa, flow rate up to 150 m^3/hr. The monitor has a system of interchangeable nozzles which provide for reliable operation at various water pressures. The barrel of the monitor can be swiveled and tilted by hydraulic jacks. The output achieved by the monitor according to field testing is as high as 40.7t/hr, including time for movement.

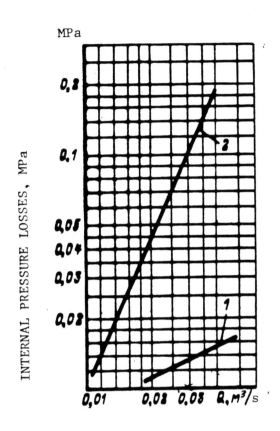

Figure 1.3. Hydraulic characteristics of monitors: 1--12GD monitor; 2-- GMDTs-3M monitor.

One direction of further development of hydraulic breaking is the creation of automated installations allowing coal faces to be worked according to programs recorded in advance.

VNIIGidrougol' Institute has developed the GMDTs-2 monitor with independent oil supply for thin seams. The remote control system of the monitor is driven by a hydraulic system using high pressure water. Operating experience has shown that hydraulic systems based on the use of process water are unreliable, with service lives of not over 20 shifts. All of this forced the Institute to abandon the process water hydraulic system and go over to an oil controlled hydraulic system.

Studies on the creation of machines which can operate unattended at the driving faces of openings and mine extraction faces are going forward in order to provide safety for operating personnel. UkrNIIGidrougl' Institute has developed the GVD-3 monitor for mining of coal in seams 0.6-1.5 m thick while unattended. The operating pressure of water is up to 10 MPa, nozzle diameter 12-17 mm for the cutting unit, 19-24 mm for the extraction unit.

It has been found that at 10 MPa pressure, a jet of water can break off a strip of coal 1 m wide and 10-12 m long through the entire thickness of a seam in 10-12 minutes. The output of hydraulic breaking is over 76 tons per hour.

The use of the new system instead of a standard monitor under the conditions of this mine has increased the productivity of labor of workers by a factor of 2.5 and reduced the consumption of timber by the same factor.

The GVD-3 unit is recommended for the conditions of hydraulic mines in the Donets and Kuznets basins for seams 0.9-1.5 m thick with coal strength factors of up to 0.7.

Cutting and extraction operations in steeply dipping, gently dipping and pitching seams are performed by the AGS unit, designed by UkrNIIGidrougl'. Coal is extracted by hydraulic breaking from a drift through a borehole 0.4 to 0.8 m in diameter, directed to the rise of the seam, for the height of one level. At the next higher drift the drilling unit is replaced by a mining unit and a strip 1 m wide and 8-12 m in length is extracted as the unit travels in the opposite direction. When steeply dipping seams are mined, the space by the face is protected from rock collapse by a pillar of coal, which is left each 5-10 m. After a column has been mined out, the unit shifts 8-12 m down the drift, and the operating cycle is repeated.

At a pressure of 8-10 MPa, the output of the AGS unit is 15 m³/hr in the drilling cycle, or up to 40 t per shift. Mining from bore holes drilled within the coal seam is performed with a water pressure of 8-9 MPa and a distance from the nozzle to the face of 13-15 m. The output of hydraulic mining with a seam thickness of 0.5 m is up to 15 t/hr, with a seam thickness of 0.7-0.8 m 30-35 t/hr. It has been found that the AGS unit can provide the maximum working safety, since it is remotely controlled.

A small remote controlled monitor has also been developed on the basis of the GMDTs-3MA monitor for mining in gently and steeply dipping seams over 0.8 m thick.

Hydromechanical machines are smaller and lighter than ordinary mining machines, since they need no loading or transport devices (for example, the PK-3 combine is about four times heavier than the LMGP-3 machine, which has the same characteristics). Studies by UkrNIIGidrougl' have demonstrated that the hydromechanical method of mining of coal can be used successfully under various conditions. Hydromechanical machines for driving of openings can increase the productivity of labor by a factor of 4 and reduce cost by a factor of 3 in comparison to driving of openings by the method of drilling and blasting.

Effective hydromechanical mining can be performed by machines with outputs of 100-150 t/hr, remotely controlled from a distance of about 50 m. The creation of hydromechanical machines which mine coal without the use of supports under remote control from development openings is a reliable means for mining of coal without personnel being present in the goaf.

A hydromechanical mining machine called the K-56, with stored-program control, has been developed for gently dipping seams.

The following units have been developed for development openings:

the AP-1 unit for driving of development openings for hydraulic stowing of rock in the goaf;

the MGPP-3 hydromechanical machine for driving of development openings through rock;

a hydromechanical cutting machine for driving of openings in seams up to 1 m thick;

14

The UPKG hydromechanical combine for driving of horizontal development openings of small cross section in thick seams.

The labor consumption of hydromechanical mining of coal is under certain conditions significantly less than the labor consumption of the ordinary technology. During hydromechanical mining, the operations related to driving of cutting openings require more labor, though less than the hydraulic-blasting method.

In the hydraulic-blasting method of mining of coal, the monitors are used only to wash down the broken coal, since the water pressure is not over 2 MPa. If the coal is broken by high pressure jets at 8-9 MPa, great ejection forces develop. The air flow under these conditions reaches 70-80 m^3/min, creating favorable conditions for good ventilation.

The hydromechanical method of mining and driving of openings could be widely used at many mines in the Donets basin; the productivity of labor would be increased by a factor of 3-3.5, costs decreased by a factor of 2.5.

Hydraulic washdown is performed using a water pressure before the nozzle of 0.8-1.9 MPa. The specific consumption of water for hydraulic washdown is 3-6 m^3/t. Hydraulic washdown of coal in cutting and mining faces should be performed with nozzles over 50 mm in diameter, providing, at this low water pressure, good throughput and minimum power consumption. Hydraulic washdown does not require high water pressure, though in order to achieve the desired water jet flight distance in the limited space of mine openings the water pressure should be at least 1 MPa, with a minimum flow rate of up to 300-400 m^3/hr.

The production rate of the washdown operation with otherwise equivalent conditions depends on the quantity of water fed to the face per unit time and is almost independent of the pressure of the jet, i.e., the power of the jet.

The length of flight of the jet during washdown has been determined experimentally:

pressure nozzle, MPa	0.5	1.0	1.5	2.0	2.5
jet flight distance, m	10.3	14.6	18.0	21.0	23.0

As the flow rate increases, the throughput of the washdown operation also increases. For example, with a flow rate of 167-270 m^3/hr for cutting faces and 147-270 m^3/hr for mining faces the production rate increases from 26 to 52.5 t/hr and 48 to 83 t/hr, respectively.

As the diameter of the nozzle and flow rate of the water increase, the power consumption of washdown decreases. With a nozzle diameter of 32 mm the power consumption is 3.3 kW·hr/t for cutting faces, 1.24 kW·hr/t for mining faces. With a nozzle diameter of 50 mm, the power consumption is 1.1 kW·hr/t for cutting faces and 0.74 kW·hr/t for mining faces.

There is no need for high pressure water for washdown of coal; only the required jet flight distance must be provided. Therefore, the pressure can be 1.0-1.6 MPa.

The use of comparatively low pressures for washdown allows pipes with smaller wall thickness and electric motors with lower power ratings to be used.

1.3. System of Working at Hydraulic Mines

Opening of a shaft mine field, i.e., driving of openings which provide access from the surface to the useful mineral in the shaft mine and allow development openings to be driven is usually performed in hydraulic mines using the same methods as in "dry" mines.

The opening of a shaft mine field may be combined with separation of the field into blocks with independent ventilation, with either independent hoisting of coal to the surface, or with hoisting combined for a group of blocks. After the field has been opened, openings are driven to allow mining operations to proceed.

The mine field may be developed in levels, panels or by a mixed method. With the level method, the mine field is divided into levels, i.e., into parts located between neighboring transportation or ventilation horizons, which are divided into extraction fields. With panel development, the shaft mine field is divided into panels. A panel extraction field usually measures 500 x 500 m. Development may be by the single panel method, meaning that there is one panel located along the strike in each seam, or by a two-panel or multipanel method.

Development openings are driven into a shaft mine field which has been opened as necessary to service the extraction operations during which actual mining of the useful mineral is performed. The main roadways, i.e., the main development openings, which are driven throughout the entire shaft mine field and are designed for servicing of the panels, are sloped at an angle sufficient to allow gravity hydraulic transport, in which the slurry has a free surface. The use of gravity transport places certain limitations on the selection of the dip angle of seams. For reliable operation of gravity transport the minimum dip angle must be 4°.

The main roadways can be driven at the same slope as in "dry" mines. In this case the slurry must be carried in pipes under pressure, such that the flow has no free surface.

An extraction field is worked at one or more faces. After a shaft mine field has been opened, the openings required to service the extraction openings are driven. The development cutting workings are driven within extraction sections; they may include crosses, roadways, etc. Extraction is performed in these extraction openings, and include breaking and delivery of the rock to the transportation equipment in the main openings. Extraction may be performed in the direction of strike (when the extraction face moves along the strike), in the direction of dip, of rise, across the strike or diagonally (when the extraction face moves at an angle of $25-65^{\circ}$ to the line of strike).

The development extraction operations within the bounds of an extraction section are coordinated in space and time by means of the system of working, which may be continuous, by columns, chambers or combined.

Systems of working with hydraulic transportation allow work to be performed without personnel present at the face and without the construction of supports in the extraction space. The small dimensions of the equipment used for mining of coal, and the lack of long walls allow hydraulic mining to be used to mine small seams under particularly difficult mining and geological conditions.

17

1.3.1. Systems of working with short mining faces with extraction in columns in the direction of dip

The most important systems of working of thin seams are systems with long columns on the rise, which are taken in lifts in the direction of dip. The coal is removed by hydraulic breaking or hydromechanical methods. This has come to be called the "polysayev" method.

The limb of a level is divided into columns by raises, which are driven from the lower accumulating roadway to the upper ventilation roadway. The columns produced are oriented on the rise or at an angle to it. The most important parameter of this system is the size of each column in the direction of strike, which varies from 10 to 26 m (Donets basin practice) with a level height of 80 to 180 m.

The coal in the columns is removed using short extraction faces from the bottom up. The line of extraction faces is stepped; the leading edge of the workings is sloped in the direction of movement of the front of extraction operations.

Ventilation and additional exits are provided by crosscuts which are driven through the columns of coal. The distance between ventilation crosses is 50-80 m.

Extraction of a single pass (figure 1.4) may be performed on one side relative to the raise (version b) or on both sides of the raise (version a).

Dead end extraction openings may be ventilated using local fans or by overall mine depression.

The accumulating roadways are driven on a wide or narrow face, the raises and crosscuts are driven using monitors.

This system of working is utilized with a seam thickness of 1-1.2 m, dip angles of 10° and more, coal strength 0.5-2.2 (Protod'yakonov strength) in mines of various gas categories. The coal is mined by hydraulic breaking or blasting plus hydraulic washdown in both cutting and extraction openings.

The advantages of this system of working include the possibility of achieving high production rates and rapid advance of the extraction face by using large numbers of raises simultaneously. The system is highly

Figure 1.4. System of working in short faces with extraction of columns in the direction of dip: a--extraction of lifts on both sides of the raise; b--extraction of lifts on one side of the raise

reliable, since problems at one face do not influence the operations at others.

The output of monitors breaking seams with strengths of 0.5, 1.2, 1.8, water flow rate 40 m^3/hr and pressure of the jet at the nozzle 5, 6, and 10 MPa has been measured as 38, 30 and 22 tons per hour, respectively.

The system of working of gently dipping seams of moderate thickness or thin seams in long columns in the direction of rise with extraction in the direction of dip in lifts ("polysaev" system of working) can assure high effectiveness of hydraulic breaking with a jet pressure of at least 10 MPa or with hydromechanical mining.

One shortcoming of this system of working is the large volume of development openings, 40-60 m per 1,000 t of coal won.

A system of working with monitor extraction from extraction crosses and block raises was introduced at "Pioner" hydraulic mine. Block raises are driven, from which extraction is performed by GMDTs-3M monitors. Each level is worked out by dividing it into sublevels 50-60 m high, in which the cutting and extraction operations are performed alternately, the upper level leading the lower level by 70 m. The width of pairs of columns is 28 m. A group of sublevels is worked in a level at the same time.

This system can decrease the time of maintenance of cutting openings, significantly improving their condition.

1.3.2. Systems of working in short extraction faces with extraction in columns in the direction of strike

This system of working has been extended in the hydraulic mines of the Kuznets basin to both gently and steeply dipping seams. Each level, bounded in the direction of dip by ventilation and accumulating roadways, is divided in the direction of strike into blocks 100-200 m in length, which can be worked in forward or reverse sequence. Raises are driven into blocks for the movement of personnel and delivery of materials. Cross cuts are driven from the raises with sufficient slope for gravity flow; therefore, the blocks are divided into columns in the direction of strike (figure 1.5).

In the hydraulic mines of the Donets basin this system has been used with seam thickness over 1 m, dip angle over 11°, coal strength 0.5-1.5 and stability of rock below average, with level heights of 60-140 m, block length in the direction of strike 100-120 m, width of a strip in the direction of rise 8-12 m, distance between ventilation cross cuts, 12-50 m.

Depending on the system of ventilation and the sequence of working and block, several variations of this system of working are possible.

Advantages of the system: assurance of high output due to simultaneous operation of large numbers of mining faces and simplicity of technology. This system of working provides a productivity of labor 1.7-2.5 times higher and a cost 25 to 37% lower than in mines with the ordinary technology.

The system of working with hydraulic washdown has no significant advantage over the use of mining machines at longwalls. The monthly productivity of labor of a worker per plan is 71.5 t (actual 61 t). With a water pressure of 7.2-7.5 MPa, the production rate of the monitor averages 30 t/hr or more for an entire lift.

In the Kuznets basin between 1971 and 1976 there was an increase in the specific share of chamber and pillar systems of working, while the share of extraction of pillars by longwalls during this time decreased. In the Donets basin during the same period of time there was also an increase in the specific share of the chamber and pillar system, though the level of extraction of columns by longwalls remained approximately the same.

1.3.3. Sublevel hydraulic breaking

Sublevel hydraulic breaking (figure 1.6) is particularly favorable in steeply dipping seams, where ordinary mechanization equipment cannot assure reliable highly effective results. Systems of working with sublevel hydraulic breaking have been successfully used with both unstable and stable surrounding rock (for example, in the Donets basin up to 80% of of mines working steeply dipping seams have stable surrounding rock).

21

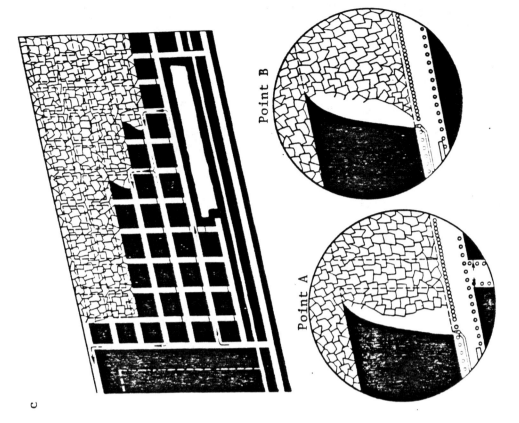

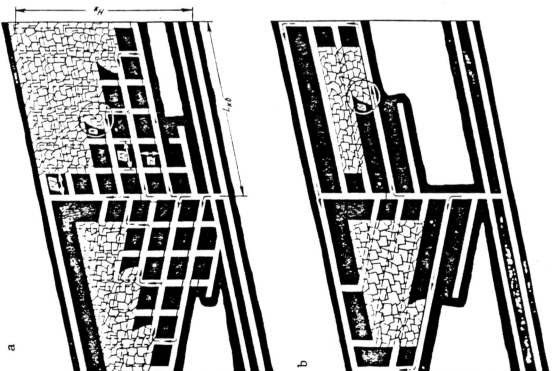

Figure 1.5. System of working in short extraction faces with extraction of columns in the direction of strike: a--ventilation in central raise; b--ventilation through flank cross cut; c--ventilation through central raise.

Sublevel hydraulic breaking is recommended for extraction of coal in seams of complex structure in the seam thickness range from 0.5 m up, with surrounding rock of any stability and gas content. The dimensions of a block in the direction of strike should be 150–300 m, in the direction of dip 100–120 m.

Coal should be extracted using the microstructure of the seams, with roof control by complete collapse without supporting the goaf, transport of the broken coal by hydraulic gravity flow transport.

Steeply dipping seams in hydraulic mines have been worked by a system with sublevel hydraulic breaking which has been found to be highly effective, though with shortcomings, since it requires a large volume of development and cutting workings. The need has arisen for improvement of the system to reduce the volume of cutting operations to the minimum and decrease the loss of coal in the goaf.

1.3.4. Specifics of working thick seams

Considerable experience has been accumulated in the Kuznets basin on the hydraulic mining of coal in seams 3.2–25 m thick with dip angles of 8 to 80°. During the past few decades, various systems of working of thick coal seams have been developed and introduced to practice:

in columns in the direction of strike with monitor extraction on one side:

in diagonal columns with combined lifts and the block development system:

in columns in the direction of rise with extraction of the coal by hydromechanical mining machines;

by sublevel hydraulic breaking using a flexible canopy.

The system of working of thick gently dipping seams in columns in the direction of strike with extraction by monitors from one side has been successfully used to mine seams 3.5 to 4.1 m in thickness with a dip angle of 8 to 18°. When this system is used, walkway raises are used both for personnel and ventilation, while the slurry is transported through slurry chutes. The raises in the extraction field are driven by K-56 MG mining machines. Extraction of coal from the pillars between extraction roadways is performed without supporting the goaf.

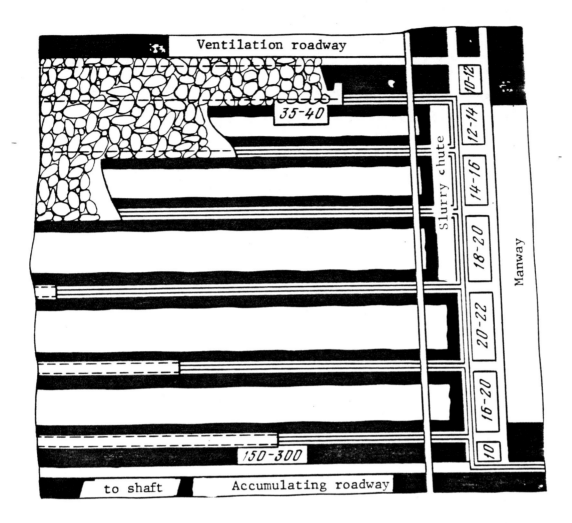

Figure 1.6. System of working with sublevel hydraulic breaking

In the Kuznets basin, a system of working in diagonal columns with combined lifts and a block scheme of development is in successful use in a seam 6.8 m thick with a dip angle of 21-25°. In this system, each level is divided into blocks with dimensions of up to 60 m in the direction of strike and 100 m in the direction of rise. As in the previous system, the entries are driven by K-56 MG mining machines. A block is divided into diagonal strips 8-10 m wide. The extraction raises are joined by bore holes 800-850 mm in diameter. The coal is won in combined lifts using GMDTs-3M machines. The openings are ventilated by

24

local ventilation fans and overall mine depression.

The mines of the Kuznets basin utilize a system of working of thick, gently dipping seams in columns in the direction of rise with extraction on one side by hydromechanical methods. A seam of moderate thickness, 3.7 m, with a dip angle of 5-12°, is being worked. As in previous systems, the entries are driven by K-56 or GPKG mining machines. Transportation of the broken coal is by gravity flow along the floor of the extraction raise.

Long operating experience has shown that this system of working yields a productivity of up to 55 tons per manshift, with operating losses of 20-22%.

When thick sloping and steeply dipping seams are worked, a system of sublevel hydraulic breaking beneath flexible canopies is used.

In this case, development of a section of a seam is undertaken by driving intermediate cross cuts at the haulage and ventilation horizons, and also by driving a cargo chute, slurry chute and ventilation raises. The seam is developed with roadways at various levels with a slope of 0.05:1 in the direction of the slurry chute. The coal is extracted with 12 GD-2 or GMDTs-3M monitors. The supports used have flexible canopies, interwoven strips of metal screen. The screen is laid out in the direction of strike and across strike of the seam in two layers at each sublevel. The work is performed so that one extraction face is in operation at any one time. Depending on the thickness of the seam, the productivity of labor is 25-40 tons per manshift.

1.3.5. Features of development operations

The following systems of working must be considered most progressive:

the system of working in columns in the direction of strike with short extraction faces and breaking of the coal by remote controlled monitors;

the same system of working, but with extraction of the coal in entries and extraction workings by self-propelled monitors;

the chamber system of working with extraction of the coal in the chambers by hydromechanical mining machines and subsequent extraction of the pillars with monitors.

In the first version, the height of a major level is taken as 250 m, the width of a block is 60 m, allowing delivery of materials, movement of personnel and assurance of stability of cross cuts. The coal is broken by monitors with a water pressure of 7.2 MPa, yielding 31 t/hr at extraction faces, 15 t/hr in entries; the loss of coal is up to 15%.

In the second version the system of working is the same as in the first but the coal is broken by monitors with operating pressures of up to 11 MPa. The coal is mined at extraction faces in long and short lifts 1 m wide under automatic programmed control. The mean output of one monitor is 37.5 t/hr.

In the third version, the coal is extracted by hydromechanical equipment. Level height and block width are both 50 m. The coal is mined with mining machines. This version has not yet been put into practice. This system of working requires complex ventilation schemes (local ventilation fans must be used) and delivery of materials and equipment is difficult.

CHAPTER 2. THE TECHNOLOGY OF HYDROTRANSPORT OF COAL AT HYDRAULIC MINES

2.1. Hydrotransport of Coal at the Face and Through Primary Openings

Two types of transport are in use in hydraulic mines: pressurized and nonpressurized (gravity) transport.

In pressurized hydrotransport a solid material mixed with water is transported through pipes. The energy necessary to move the slurry is supplied by coal pumps or hydraulic elevators (jet pumps).

The openings through which the pipes are laid do not need to be sloped as is necessary for gravity hydrotransport, since the coal pumps transport the coal through horizontal, sloping or even vertical openings. If the head created by one coal pump is insufficient, two or more pumps are connected in series.

Pressurized hydrotransport in a mine requires additional capital investment for the construction of equipment.

The hydrotransport of coal within a section is by gravity through flumes laid in roadways. The costs of gravity transport are not great, and when it operates normally, no special personnel are required to clean out the flumes or roadways.

Hydrotransport of coal within a section almost completely eliminates the expenditure of labor in transport during hydraulic mining.

According to VNIIGidrougl', nonpressurized hydrotransport costs one-third as much as conveyor transport and is half as expensive as electric rail haulage.

The technological system calls for gravity hydrotransport of all slurry to a chamber by a shaft and clarification of slime water before hydraulic hoisting.

A portion of the slime water is removed from the slurry using a water separator, then the thickened slurry is transferred to the hydraulic hoist chamber, the slime water is sent to the water collector of the pumping plant. The clarified water is returned to the surface by pumps, the thickened slime is transported further by coal pumps.

Gravity transport is normally used within the limits of extraction fields, frequently in main openings for transportation of slurry to the hydraulic hoist chamber.

The design of gravity hydrotransport can be reduced to selection of the slope of the flumes and determination of their dimensions. Reliable hydrotransport of coal requires that a slope of 0.05:1 be maintained, while the dimensions of the flumes must be determined as a function of the flow rate of slurry (table 2.1).

Flow rate of slurry, m³/hr	Type	Overall width, mm		Height, mm	Cross section, m²
		Top	Bottom		
100	I	280	200	250	0.06
200-300	II	380	300	300	0.10
500-600	III	500	400	300	0.13
1000-1200	IV	600	550	300	0.17

One shortcoming of gravity hydrotransport is that it makes the use of rail transport for delivery of materials and personnel more difficult.

In hydraulic mines, nonpressurized and pressurized flows of water and slurry are joined by means of buffer containers. Water may be recirculated from the face to the coal pumping station in a section. In the crushing and sorting section, a portion of the slime water is separated from the slurry on screens, then the water from beneath the screen is sent to a settling tank, from which it flows by gravity to the pumps. The thickened slime is sent to a regulating container at the coal pumping station of the section, then transferred by coal pumps to the main hydraulic hoist chamber. The design of recirculation water supply systems for hydraulic mines must consider the unsteady nature of arrival of slurry and solids from the faces to the main hydraulic hoist chamber. Centrifugal machines can be easily adapted to various operating conditions, but frequent shutdown and startup of coal pumps requires additional expenditure of water for washing and refilling the system, and also causes the machinery to operate under unsteady conditions, resulting in more rapid failure.

Type 8 MKD monorail systems can be used as supplementary transport in hydraulic mines where the transportation distance is at least 1200 m, maximum permissible upward slope 8°. RKD cableways are also used. The cost of operation of monorail systems per ton of coal won is lower than the usual system by a factor of 2-2.5.

The cost of hydrotransport within a section (including replacement of won flumes and cleaning of roadways) is not over 2.5% of the cost of mining the coal, i.e., approximately one-third the cost of the usual technology.

The selection of an efficient type of hydraulic system for a hydraulic mine of a given capacity under specific mining and geological conditions is based on technical and economic calculation, considering capital investments, reliability, safety of operation of the system and its operational cost. In selecting a version one must keep in mind that the gravity hydrotransport system has a long operating life and significant capacity for expanded throughput.

2.2. Preparation of Coal for Hydrotransport

Reliable operation of a hydrotransport system can be achieved only by proper organization of the preparation of the slurry. Before the slurry is input to the coal pump, outsized lumps and foreign objects, particularly metal, must be removed. The removal of the chips which accumulate in collectors must be mechanized.

Before hydrotransport, the coal is crushed to a maximum particle diameter of 50-60 mm, since otherwise it cannot be fed through centrifugal pumps with water. The coal must also be prepared for gravity hydrotransport.

The "Balmer-South" hydraulic mine (in Canada) uses feeder-breakers which are installed ahead of the flumes. The crusher consists of a number of discs attached to a shaft, spaced at a distance of 100 mm. The coal is crushed by teeth with hard alloy tips. Coal is fed to the installation by a two-chain conveyor. The use of feeder-breakers has achieved positive results, since a steady flow of slurry is assured and the movement of the heavy monitor is mechanized by the same device.

The process of crushing before pressurized hydrotransport may consist of preliminary and check screening. Crushing may include one, two or more stages; the number of stages depends on the initial and required final particle size of the material being crushed. Preliminary screening is used to avoid crushing anything which does not need it. The economic expediency of additional capital investment required for the preliminary screening devices can be demonstrated if the amount of material which

29

passes through the screen is comparatively great; if little material passes through the screen, preliminary screening does not make sense.

The expediency of preliminary screening should be demonstrated for each specific case.

In selecting a type of crusher, it must be considered that the throughput of a cone crusher is 2.5 to 3 times greater than the throughput of a jaw crusher of the same intake width; therefore, where the required throughput is low, the cone crusher may be underutilized. Cone crushers are the most common type used in hydraulic mines.

DKU hammer crushers with rock throughputs of 120-180 t/hr, installed together with grizzlies, have become increasingly popular recently in hydraulic mines.

The experience of their use indicates that hammer crushers are highly reliable in comparison to tooth, cone and pick crushers, but that they crush the coal overfine (content of particles less than 1 mm in diameter increased by 2.3%).

Hammer crushers are distinguished by their low power consumption, simple design and low weight. The West German firm "Gerb Hischman" produces a series of hammer crushers of welded design with two hammer plates lined with wear-resistant materials. Access to the plates is simple, allowing inspection and quick replacement of lining. The hammers are made of high alloy steel and attached to the rotating discs of a rotor. The hammer shape depends on their purpose (primary crushing or fine grinding). They are made of one or two elements -- a head of wear-resistant steel and a middle section. When moist materials are crushed, hot gas can be fed through the crusher to prevent the formation of lumps.

The advantages of cone crushers are: rigidity, high reliability, low specific power consumption and high degree of crushing (particle size reduction ratio in crushers varies from 1:20 to 1:50). The body and side parts are lined with steel plates and installed on a frame of rolled shapes with high wear resistance, impact elements are made of cast wear-resistant steel. The rotor is carried by self-centering roller bearings. The teeth are self-sharpening due to their reverse profile.

A hydraulic device with a preventer or reverse valve is used to

30

remove objects which cannot be crushed. When the reverse valve opens, the rotating element rotates backward to its initial position, until the hard object can pass through. The crusher is dynamically balanced and can be installed on any horizontal surface.

2.3. Technological Hydraulic Hoisting Systems

One of the most important problems for hydraulic mines is the selection of a reliable type of hydraulic hoist. The most commonly used hydraulic hoist systems are: coal pumps, systems with feeding apparatus and air lifts. These types of hydraulic hoists have been tested under industrial conditions and each has its own area of application.

Coal pump hydraulic hoists are the most common type of slurry hoist system (figure 2.1-2.3).

Coal pumps currently manufactured create maximum pressures of up to 32 MPa: multistage operation of coal pumps is also possible.

A coal pump hydraulic hoist has the following features:

possibility of transporting slurry through horizontal, sloped and vertical pipes; it is therefore possible to feed slurry from the mine to the beneficiation plant or consumer with any configuration of the pipe path;

no need to construct an additional water drainage installation to pump out water which flows into the mine when coal mining operations are interrupted; the specific power consumption of a coal pump hydraulic hoist is lower than that of other systems;

coal pump installations can be easily automated.

The shortcomings of coal pump hydraulic hoisting are as follows: short service life of coal pump installations (not over 4,000 hours); when several powerful coal pumps are installed on the same level with large electric motors, additional electric power must be expended to provide forced ventilation of the coal pumping chamber; worn parts and units of coal pumps of great size and weight must be frequently renewed, significantly increasing operating costs.

Coal pump hydraulic hoisting is expensive in terms of operating cost when unreliable equipment is used.

In spite of these shortcomings, coal pump hydraulic hoisting is the most commonly used type both in our country and abroad. Of ten operating hydraulic mines in the USSR, eight are equipped with coal pump hydraulic hoists.

31

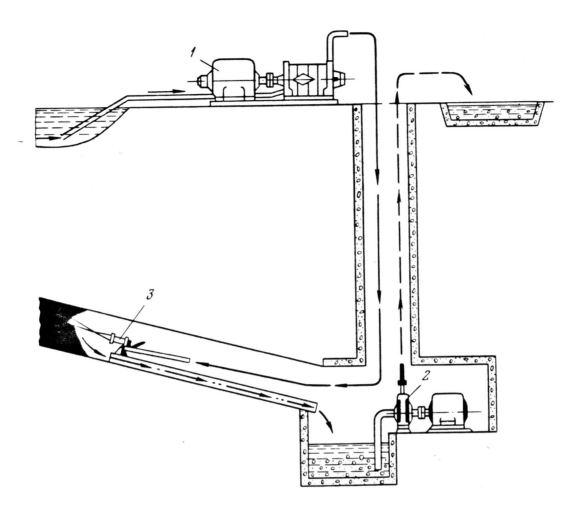

Figure 2.1. Coal pump hydraulic hoist with pump installed at the surface:
1--high-head pump; 2--coal pump; 3--monitor

Hydraulic hoisting with loading apparatus (figure 2.4) has been through
commercial testing in our country and has been tested under various
conditions; hydraulic hoist systems with feeding apparatus have been in
use abroad for some time. The advantages of this type of hydraulic
hoisting are as follows: minimum crushing of the coal in comparison with
coal pump and air lift hydraulic hoists; possibility of feeding slurry
directly to the beneficiation plant or to a large consumer through a
pipeline of any configuration; comparatively low power consumption.

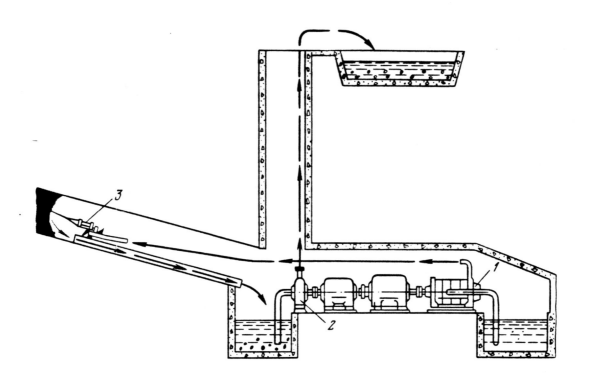

Figure 2.2. Coal pump hydraulic hoist with pump installed in the mine:
1--high head pump; 2--coal pump; 3--monitor

Hydraulic hoisting with feeding apparatus has not yet been used in
hydraulic mines in the USSR on a commercial scale. Problems of providing the necessary strength of high capacity apparatus for operation
at high pressures (over 30 MPa) have not yet been solved. Normal
operation of low capacity feeding apparatus requires the creation of
reliable valves and other pressure control devices suitable for frequent
operation. The experience which has been accumulated is now sufficient
for these devices to begin to be applied under various hydraulic hoisting
conditions at the present time.

Air lift hydraulic hoisting has the following advantages for hydraulic
mine conditions (figure 2.5):

the primary equipment (compressor station) is installed on the surface,
no equipment with moving and rotating parts needs to be installed underground, there is no need for additional expensive ventilation of shaft
openings;

the need for high pressure equipment is minimized;

mine openings near the shaft can be of minimum dimensions.

However, air lift hydraulic hoisting consumes the greatest amount of power of all types of hydraulic hoisting and is suitable only for vertical or sloping pipes; it cannot always be used for direct transportation of slurry to a beneficiation plant or consumer. Therefore, coal pumps or feeding apparatus must be used in addition to air lifts.

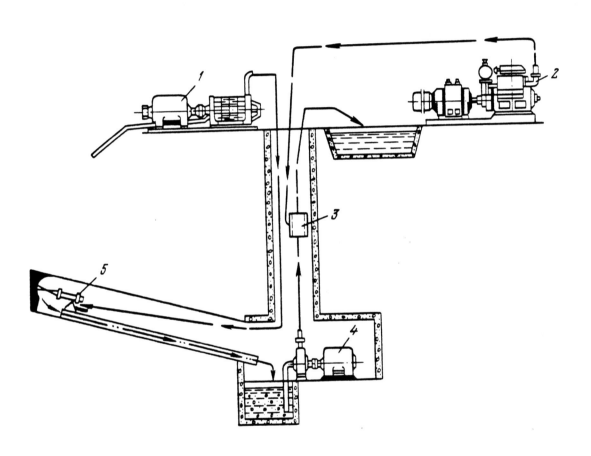

Figure 2.3. Combined hydraulic hoisting system: 1--high-head pump; 2--compressor; 3--air mixer; 4--coal pump; 5--monitor

Air lifts, like coal pumps, crush the coal overfine, which limits the area of their use primarily to coal to be used for power generation.

The power consumption of coal pump hydraulic hoisting is significantly less than the power consumption of air lift hydraulic hoisting; the degree to which the coal is crushed in these types of hoists is comparable. The reliability of operation of air lift hoisting, using industrially manufactured equipment and units, is equal to that of coal pump hoisting; therefore, depending on conditions, either of these types of hydraulic hoist equipment can be recommended. The final selection must be based on technical and economic analysis.

2.4. Hydraulic Mine Water System

During mining operations, pumping plants on the surface feed water into the hydraulic mine. Extraction and haulage of one cubic meter of rock mass, operating experience has shown, requires a significant quantity of process water, the exact amount of which depends on the physical and mechanical properties of the rock involved. The effectiveness of hydraulic mining can be increased by decreasing the consumption of process water if steps are taken to increase the concentration of the slurry delivered to the hydrotransport units.

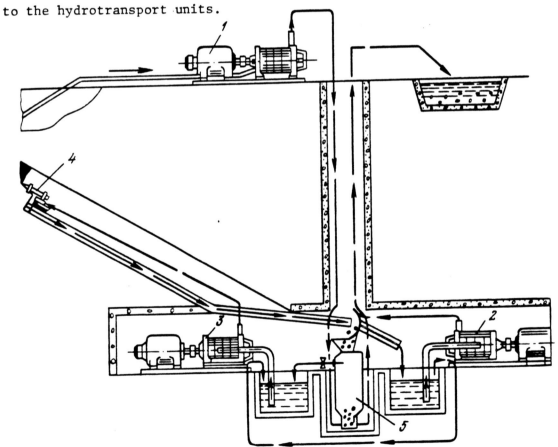

Figure 2.4. Hydraulic hoist system with loading apparatus: 1-3--pump; 4--monitor; 5--feeding apparatus.

Present hydraulic mines are supplied with water with high-head pumps. High pressure water is fed from the surface to monitors. The water supply system consumes large amounts of power and the distribution system is quite complex.

Systems of feeding water to the faces using medium pressure pumps have been developed, and the necessary high pressure water for hydraulic breaking is obtained by the use of pressure boosters such as those developed by VNIIGidrougl'. This allows the use of less expensive pipes and fittings, simplifies the servicing of pumping units and reduces the cost of maintaining the power system.

Hydraulic mines are large consumers of water. Usually, a closed water supply cycle is used, i.e., the process water is repeatedly reused. At least 15-20% of the total water which flows through the system must be supplied from outside the system to its reservoirs. Furthermore, process water must be clarified, purified of mechanical impurities and abrasive particles.

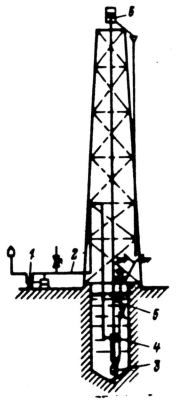

Figure 2.5. Air lift hydraulic hoist system: 1--air pump; 2--air line; 3--intake device; 4--mixer; 5--hoist pipe; 6--air separator

In designing the water supply systems for hydraulic mines, the basis used is the daily output and operating conditions at the face, calculated so as to maintain a steady water flow rate. Maximum water consumption occurs when all consumers are on line at once, but the probability of this is very low. It is therefore assumed in design calculations that the monitors in entries and at production faces will operate not over three hours per shift, the remainder of the time being occupied by development and closeout operations. Existing monitors require a flow rate of 900-2000 m^3/hr of water at a pressure of 2-10 MPa for normal operation.

Hydraulic mining of many types of coal requires pumps which can develop pressures of 10-16 MPa. Sometimes the pumps in hydraulic mines operate in a closed cycle; therefore, they must be designed to operate with a certain quantity of solid impurities in the water. Small centrifugal pumps are most suitable for this purpose, have low mass and high efficiency. Series produced low head pumps manufactured by domestic industry for clear water at pressures of up to 10 MPa and flow rates of up to 1080 m^3/hr are used in hydraulic mines. MS series pumps manufactured by the Yasnogorsk machine plant are most widely used.

Pressure control devices such as slide valves and check valves are important for successful operation of hydraulic mines. Many designs of slide valves and check valves operate satisfactorily with clear water but quite poorly with slurry.

One means for increasing the effectiveness of operation of hydraulic mines is to provide for partial recirculation of the water within sections, i.e., return of a portion of the process water from the standby and regulation containers of the coal pumping chambers of sections to the faces. The use of recirculation can reduce the amount of water which must be fed into the mine by 40-50% and increase the concentration of slurry hoisted out of the mine by a factor of 1.6-1.7.

The recirculation of process water within the mine can be organized by two main systems:
1) section face -- central hydraulic hoist chamber -- section face;
2) section face -- section coal pumping chamber -- section face.

In system 1, water is fed in the quantity necessary for mining of the seam from the hydraulic hoist chamber. The slurry is fed to the section coal pumping chambers, from which it goes on to the hydraulic hoist chamber, and is then delivered by coal pumps to receiving devices. at the beneficiation plant. The slurry fed to the beneficiation plant can have a higher concentration than in the sections in the mine. This system can be used with gravity hydrotransport from the sections to the hydraulic hoist, with the coal extracted by either mechanical or hydromechanical equipment.

Another system is also possible, in which the water is recirculated within the mine from the faces to the coal pumping chamber in a section. This system can be used within the section with subsequent pressurized hydrotransport from the coal pumping chambers in the sections to the hydraulic hoist chamber. Systems with recirculation of water within the mine can be used if high pressure water is not needed, for example, with hydraulic washdown of the loosened mass of coal during operation of hydromechanical devices. This does not require great clarification of the recycled water, nor does it require installation of high power devices in the mine.

If recirculation systems are used with hydraulic breaking of the coal, it is necessary to clarify the high pressure water quite thoroughly, and also to install powerful hydrotransport devices in the mine, which is quite expensive and difficult under mine conditions.

If the water pressure needed for breaking of the coal is comparatively low, water recirculation within the mine may be economically justified.

Recirculation systems should be so constructed that the water supply cycle involves the smallest possible number of pipes. The problem arises of underground purification of the water, since process water containing large quantities of suspended solid particles causes rapid wear of equipment. It may not be suitable to construct additional containers and pumps.

For this reason, water recirculation systems have not been very widespread.

2.5. Dewatering of Coal at Hydraulic Mines

The technology of dewatering and drying of coal after hydrotransport consists of three operations: thickening of the slurry, dewatering of the coal by mechanical methods and thermal drying.

Preliminary dewatering of coal to a moisture content of 30% is performed on a conical screen (CS) designed by UkrNIIGidrougl', with a throughput of up to 140 tons of solid material per hour.

The slurry is thickened in cylindrical thickeners with tank diameters of up to 100 m and depths at the center of up to 7 m and thickening areas of 3.1 to 7850 m^2. Thickeners with central drive have maximum tank diameters of 18 m, while peripheral drive thickeners may be up to 100 m in diameter.

The throughput of thickeners depends on the properties of the incoming slurry. The design of thickeners varies: conical thickeners, pyramidal settling tanks, settling tanks with flat bottoms and outdoor settling basins.

Slurries are thickened in a thickener with a sediment stowing device developed by UkrNIIugleobogashcheniye Institute [4].

This thickener can be used instead of radial thickeners for high-ash slime, producing almost clear water (3 g/l) and highly concentrated sediment.

A radial thickener with a central drive and hyperboloid bottom is now being industrially produced, designed for a throughput of 1600 tons of solid material per shift.

Slurries may also be thickened in hydrocyclones (table 2.2).

Table 2.2

Type and Size	Hydrocyclone		
	G-6	G-9	G-12
Throughput of slurry, m^3/hr:			
at 10 m head	300	450	–
at 20 m head	–	650	1000
Diameter of fitting for output of thickened product, mm		50–120	
Mass, kg	890	1223	2654

Fine coal concentrates with particle sizes of less than 6-12 mm are usually dewatered in two passes on screens, nonmoving slot screens and filtering centrifuges. Fine coal concentrates are dewatered on light vibrating screens with throughputs of up to 300 t/hr as well.

- The dewatering of finely ground materials and slimes is most frequently performed in cylindrical thickeners or pyramidal settling tanks and cones, and also on vacuum filters or filter presses.

Finely ground materials may be dewatered in one pass using a type NOGSh centrifuge with a throughput of up to 300 m^3 of slurry per hour, up to 50 tons of coal per hour, with a particle size of the dewatered material of 0-1 or 0.13 mm. The moisture content of the dewatered slime is as follows as a function of the content of the fine fraction: 0-25 mm fraction up to 36%, 0-13 mm fraction up to 20-27%.

Vibration and helical filtering centrifuges with throughputs of up to 150 tons of coal per hour with an initial coal particle size of 0-13 (25) mm, moisture content at the initial coal of 25-30% and moisture content of the dewatered sediment of 5-9% are used for the second dewatering operation (with a content of the 0-1 mm fraction in the feed of not over 25%).

Vacuum filters for fine coal slime and flotation coal concentrate are designed as a function of the throughput of dry material. Dewatering of fine classes of slurry for hydrotransport may be performed by centrifuge. A design has been developed for a vibration centrifuge using filtering elements and circular vibrations. The TsKS-1450 centrifuge has a dewatering surface of 4 m^2 with the same geometric dimensions as the TsVP-1100 centrifuge, the filtration surface of which is 2 m^2.

The VG-2K two-stage vibration centrifuge has a filtering screen with a slot width of 0.5-0.75 mm, allowing direct separation of small particles from the sediment. The throughput of this centrifuge is 90 tons of solids per hour or 400 m^3 of slurry per hour.

The NOGSh-1100A centrifuge with optimal diameter 1100 mm is now being industrially produced. The NOGSh-1100F precipitation filtering centrifuge is being prepared for manufacture; this device can produce a concentrate with a moisture content 10% lower than that produced in ordinary centrifuges, or 2-4% lower than the product of vacuum filters.

Since 1975, vibration filtering centrifuges TsVN-1120 have been produced, allowing the moisture content of the sediment to be decreased by 1.5-2%. This centrifuge has a throughput of initial coal of up to 110 tons per hour with a particle size of the input coal of 0-13 mm, initial moisture content 20-30%, moisture content of the dewatered coal with up to 10% 0-0.5 mm fraction not over 8.5%.

Further dewatering of fine fractions of coal, slime and flotation concentrate with simultaneous clarification of the recycled water can be performed on the NOGSh-1120F centrifuge. Its features include the presence of a filtering rotor in the area where the sediment is unloaded, a cylindrical-conical sediment rotor allowing the moisture content of the dewatered sediment to be decreased while decreasing loss of coal with the fugate. The NOGSh-1120F centrifuge has a throughput of 120-130 m^3 of slurry per hour.

UkrNIIugleobogashcheniye and Gipromashugleobogashcheniye have developed the S-10UR cylindrical thickener. The charging device receives a flow of slurry which, being rotated, is uniformly distributed in the horizontal plane of the cross section of the thickener. The force of gravity causes solid particles to settle into the lower conical portion of the thickener. In order to accelerate sedimentation of the particles, the loose layer of particles is disturbed by an agitator. It is thought that the throughput of this thickener will be 300 m^3 of slurry per hour, 15 tons* of solids per hour.

The effectiveness of dewatering is most greatly influenced by the particle size distribution of the feed material: the moisture content of the sediment is greatly increased when the 0-0.5 mm fraction is present, each 5% of this fraction increasing the moisture content of the sediment by 1%.

Dewatering equipment is selected as a function of the moisture content and ash content of the initial product and the requirements placed on the end product. The moisture content of materials for hydrotransportation, the so-called total moisture, includes gravitational, capillary and

---

* Note (Probably should be 150 tons - Ed.)

hygroscopic moisture. Hygroscopic moisture is not removed in dewatering.

In the USA, coal is dewatered on screens and in centrifuges. The coal is first dewatered on nonmoving screens, with subsequent dewatering on two-screen vibrating machines with apertures in the bottom screen of 0.25-0.50 mm. Centrifugres of the following types predominate: filtering, vertical and horizontal centrifuges with throughputs of up to 250 t/hr.

Various types of filters are used, primarily disc filters. The moisture content of the concentrate after filtering is about 20%, while the tailings contain up to 30% moisture. The process of dewatering can be improved by further drying the sediment with superheated steam.

Final dewatering is performed in thermal dryers, where fine coal, flotation concentrates and slime are dried. All new and reconstructed beneficiation plants use fluidized bed dryers with throughputs of 3-70 tons per hour.

The diagram of dewatering of run-of-mine coal shown in figure 2.6 calls for reception of the slurry in a separator and dewatering on a "Luganets" screen (throughput up to 1000 t/hr). Material with particle diameter over 13 mm and moisture content 7-8% is sent further to an accumulating hopper, then to the consumer. The 1.5-13 mm fraction is treated in the preliminary dewatering apparatus and a vibrating horizontal filtering centrifuge with a throughput of 320 t/hr. The slurry containing the 1.5-0 mm fraction is thickened in a thickener with a hyperboloid bottom, then dewatered on a disc vacuum filter. After dewatering, the coal may be sent to accumulating hoppers or to the dryer.

A diagram of the process of dewatering the 3-0 mm fraction of coal after hydrotransport is shown in figure 2.7. As in the previous system, the slurry is received and distributed in the slurry separator, then fed to a low-pressure hydrocyclone where the slurry is thickened and the coal is classified with a separation at 0.5 (or 1.0) mm particle size. The thickened hydrocyclone product (over 200 g/1) is sent to a horizontal screw-type precipitation centrifuge. The moisture content of the dewatered product is 18-24%.

When required for final dewatering the material may be passed through a second stage of centrifugation using lubricants (kerosene, fuel oil and

surfactants). In this case the moisture content of the dewatered
product is 16-18%. The centrifugate and hydrocyclone drain product
are thickened to 200-400 g/l, then finally dewatered in a two-stage
centrifugation process.

Figure 2.8 shows a diagram of the dewatering of coal before it is fed
to the boilers of a thermal electric power plant for combustion. The
slurry, containing coal in the 1-0 or 0.5-0 mm fraction, is sent to the
intake sump, from which it is transferred by pumps to thickeners with
hyperboloid bottoms. After thickening, the density of the slurry is
200-500 g/l. In the second stage, dewatering is continued to 20-23% in
a disc vacuum filter. Throughput is increased by the addition of
surfactants to the disc vacuum filter bath. The addition of 10-20 g/t
surfactant increases the throughput of the vacuum filter by 30 to 50%.

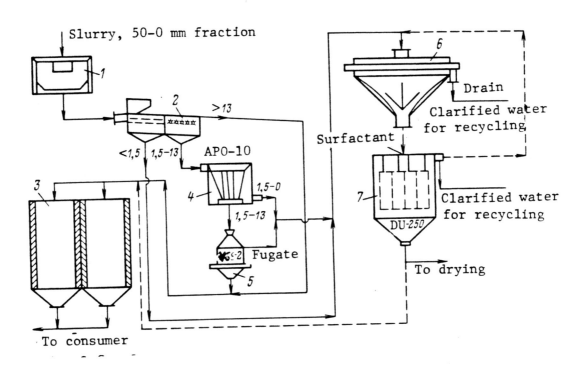

Figure 2.6. Diagram of dewatering of run-of-mine coal after hydrotransport:
1--slurry separator; 2--screen; 3--accumulating hopper; 4--preliminary
dewatering apparatus; 5--horizontal vibrating filtering centrifuge;
6--thickener with hyperboloid bottom; 7--disc vacuum filter.

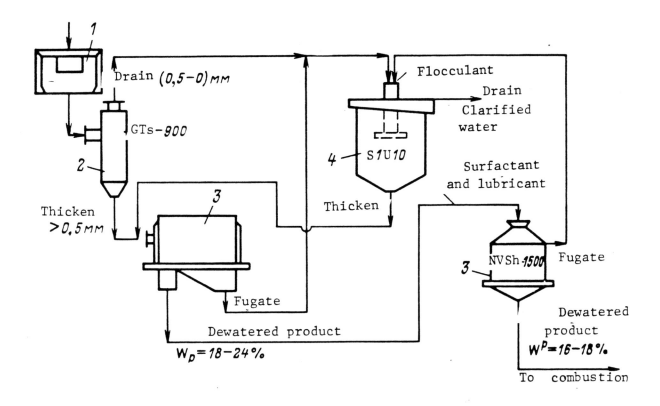

Figure 2.7. Diagram of dewatering of 0.3 mm fraction: 1--slurry separator;
2--low pressure hydrocyclone; 3--centrifuges; 4--thickener with sediment
compactor.

If the slurry contains a large number of particles in the micron
fraction (0.074 mm) it is recommended that superheated steam be fed into
the drying zone in order to increase the dewatering rate. The material
is ground to a maximum particle size of 0.2 mm before combustion.

In order to increase the efficiency of the boiler unit, the I.I.
Polzunov Institute (TsKTI) recommends that stack gases from the boiler
be fed into the drum of the ball mill. This system, which is called a
closed system, is easily automated and represents no danger of explosion.

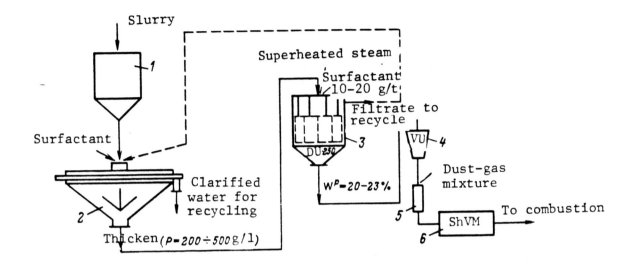

Figure 2.8. Diagram of dewatering of coal transported to thermal electric power plant: 1--receiving hopper; 2--thickener with hyperboloid bottom; 3--disc vacuum filter; 4--coal hopper; 5--mixer (coal and dust-gas mixture); 6--drum ball mill.

CHAPTER 3. TECHNOLOGY OF LONG RANGE HYDROTRANSPORT OF COAL

3.1. Basic Technological System

- The basic technological system for hydrotransport of coal over long distances consists of three main units: the slurry preparation unit, located at the fuel supply, the slurry pumping plants and the unit which receives the slurry at the consumer.

The fuel suppliers are hydraulic mines, mines using the standard technology of coal mining or open-pit mines.

In the presently existing system of hydrotransport (figure 3.1, line 1) coal which is won in shaft mines using the standard mining technology or coal from open-pit mines is fed into a device which loads it into railroad cars and it is then transported by rail to the consumer, where the coal is ground and sent to the thermal electric power plant boiler units.

In the system shown in line 2 of figure 3.1, coal from a shaft or open-pit mine is mixed with water, then the slurry is transported by pumps and pipes to the consumer, where the coal is dewatered, dried, ground to the required particle size and utilized in the boilers.

Line 3 of figure 3.1 shows a technological system in which, in contrast to the earlier systems, the coal supplier performs fine grinding, then prepares the slurry, which is transported to the electric power plant where it is thickened and transported to the boilers for immediate combustion without preliminary dewatering and drying.

The next technological system (line 4, figure 3.1) differs from system 3 in that the slurry is sent from the supplier to a device for wet grinding of the coal, then to a device for preparation of the proper mixture, which is burned by the consumer after hydrotransport.

Line 5 of figure 3.1 shows a technological system in which the slurry does not undergo further processing at the supplier, but rather is hydro-transported to the consumer, where the coal is dewatered, dried, ground and used in dry form.

The next two technological systems (lines 6, 7 of figure 3.1) differ only in the methods of utilization of the fuel by the consumer. System 6 calls for wet grinding of the coal, thickening of the water-coal mixture, which is then fed into the boiler for immediate combustion.

- In the final technological system (line 7 of figure 3.1) the coal is dewatered and dried at the consumer's location, then fed to the boilers of the thermal electric power plant.

Hydrotransport may be performed by various means in any of these technological systems -- with or without intermediate buffer storage containers. When intermediate containers are used there is a break between the pumping plants and the pipeline. All of the slurry is sent to the reservoir, from which it is transported to the next pumping plant.

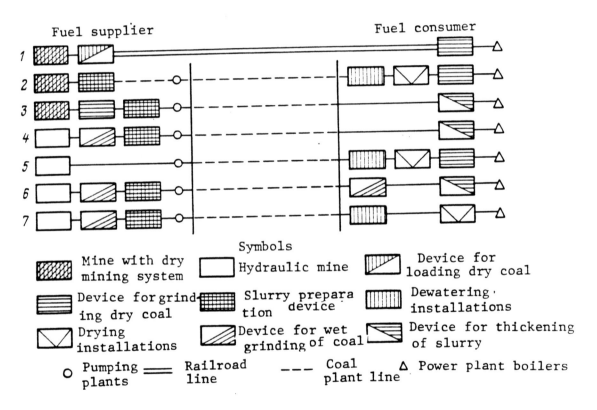

Figure 3.1. Technological systems for feeding coal to thermal electric power plants: 1-7, various versions.

The technological system is possible in which the slurry to be transported moves directly to the pump, bypassing any reservoir. During certain periods, when it is necessary to match operations, the slurry may be fed periodically into reservoirs.

The pump-to-pump transportation system is most common, though it is suitable only for centrifugal pumps, as it requires the least capital investment for construction of the slurry line.

There have been two pipelines for delivery of coal to consumers from "Yubileynaya" mine and "Inskaya" mine No. 2 in operation in the USSR since 1966-1967.

These installations are technological systems consisting of a hydraulic mine, hydraulic transport installation and installations for fuel utilization including dewatering and drying of coal. Hydraulic transport allows the mine-supplier to be united with the beneficiation plant and thermal electric power plant into a single technological process, thus decreasing the loss of coal in transportation and improving the economy of operation not only of the coal enterprise but of the large scale consumer as well.

The technological systems of power systems and fuel-metallurgical systems in which coal is fed to beneficiation plants, and to a state regional electric power plant, are approximately the same (figure 3.2). Sequentially operating coal pumps are used for hydrotransport of coal at the hydraulic mine, hydraulic hoisting of the coal to the surface and delivery to the consumer.

The particle size of the coal to be transported is determined by the cross sections of the coal pump impeller sections, and is usually not over 50-60 mm. 350 mm diameter pipe has been laid for the transportation of coal, allowing up to 2000 tons of coal per day to be transported with relatively low slurry concentration. At "Inskaya" mine No. 2, two pipelines have been planned for the transportation of 4000 tons of coal per day; at "Yubileynaya" hydraulic mine, two and three pipelines have been planned for the transportation of 4 and $6 \cdot 10^3$ tons of coal per day. 100% standby redundancy of slurry lines and water lines is planned, with

48

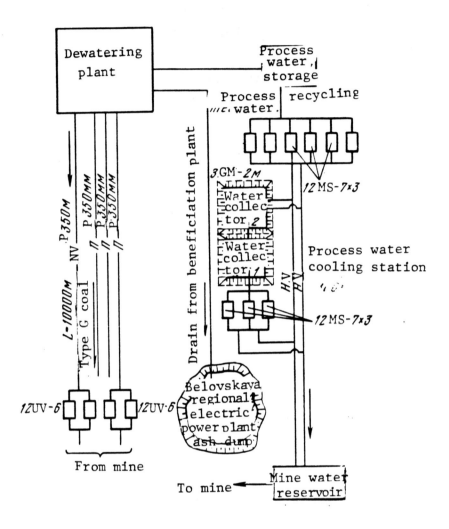

Figure 3.2. Diagram of main hydrotransport line at "Inskaya" mine No. 2

three coal pumps for each slurry line, one of which will be in operation, one on standby and one in repair.

The experience of operation of the hydraulic mines of the Donets basin and Kuznets basin has caused VNIIGidrougol' Institute to recommend the use of systems consisting of two rather than three machines.

Operating experience indicates that effective operation of hydrotransport installations depends primarily on the quality of preparation of the slurry before hydrotransport. The preparation of slurry includes operations involved in classification of the mass before crushing and removal of metal objects and wood chips from the coal. Classification and crushing are designed to achieve the largest possible diameter of lumps of solid material, as determined by the clearance in the coal pump channels.

Transportation of coal with smaller weighted mean particle diameter requires less power, causes less wear of pipes, equipment, and less further grinding of the coal. Therefore, the coal to be used by thermal electric power plants or coal for metallurgical production should be crushed to the 0-13 and 0-25 mm fractions before transportation over distances of 10-20 km. For long distance transportation, the weighted mean particle diameter of the coal should not exceed 1-2 mm. This allows the energy cost of transportation to be reduced and increases the concentration of the slurry by weight, to as much as 50% or more.

There is significant room for increasing the effectiveness of operation of existing systems, which allows us to recommend the creation of such systems as a primary area of concern for the future development of hydromechanization of the coal industry.

In main coal pipelines in the Kuznets basin, pipelines are rotated through angles of 90-120 or even 180°, the slurry transportation speed is 3.2 m/s, and particle size distribution is as follows:

particle diameter, mm	0-1	1-3	3-6	6-13	10-25	+25
content, %	30	20	20	10	15	5

Coal pipelines in use in the USSR for transportation of coal from mines to coke chemical plants and thermal electric power plants have not achieved the best possible economic results from the standpoint of hydrotransport, since the concentration of the slurry transported in them is still low, and the labor consumption of hydrotransport is high due to the excess quantity of water. The capital investment required for large diameter pipes is also too great. Nevertheless, hydrotransport is used, since it does not involve the extra cost of loading and unloading operations. As a result, the net cost is one-half to two-thirds the cost of rail transport, particularly considering that the cost of rail transport over local approach lines is higher than that of rail transport over main lines.

50

Hydrotransport systems in the Kuznets basin have achieved good economic results because the technology of hydraulic mining, supplemented by the similar process of hydrotransport within the mine, hoisting of coal to the surface and its transportation to the consumer are quite effective when combined, easily mechanized and automated, and do not require the additional cost of loading and unloading operations.

Long range hydrotransport systems are planned considering the possibility of mining, transport and utilization of coal in the form of finely dispersed water-coal slurries (suspensions). A plan for a fuel and power complex utilizing hydrotransport of coal, drawn up by the Khar'kov division of Teploelektroproyekt Institute, involving transportation of coal from a hydraulic mine to the Novo-Dneprovskaya regional electric power plant 426 km distant, with a total throughput of four million tons of coal per year, has demonstrated the excellent economy of such a system and the possibility of decreasing the cost of electric power by 20% of the cost with traditional types of transportation and combustion of coal.

The first large commercial coal pipeline 173 km in length was put into operation in 1957 in Ohio (USA). This pipeline was 254 mm in diameter and was used to transport 1.3 million tons of coal each year. The utilization of hydrotransport alone decreased the cost of transportation of coal by 43%, and the railroad companies were twice forced to decrease their tariffs.

The experience which has been accumulated in the operation of long distance coal transport systems has allowed full scale commercial operations to be started. In the USA in 1970 what is now the world's largest coal pipeline, 440 km in length, with a throughput of over 5 million tons of coal per year, was put in operation, transporting coal from the Black Mesa mine to the Mojave 1500 MW power plant on the Colorado River near Bull Head City, Arizona. A slurry preparation unit is located near the mine. The incoming coal, in the 50-0 mm particle size range, is crushed by hammer crushers to 10 mm maximum particle diameter, then further crushed in rod mills to a maximum particle diameter of 1.2 mm.

51

Since the thermal electric power plant burns finely ground coal, the energy expenditure for grinding of the coal does not represent an additional cost of hydrotransport.

After grinding, the coal, mixed with water, is sent to settling tanks, from which the slurry is transferred by centrifugal pumps to collecting reservoirs. One of the reservoirs is filled with slurry for two hours, while the consistency of the slurry is adjusted to the required level of 50% by weight in the other reservoir, from which the slurry is fed to a third reservoir, then into the main pipeline. Final checking of concentration is performed before the slurry is fed to the transport pumps.

The slurry is transported in steel pipes 457 mm in diameter for the first 413 km and 366 mm in diameter over the remainder of the distance, with pipe wall thickness varying from 5.5 to 12 mm. The route of the pipeline, shown in figure 3.3, crosses five mountain chains and two rivers, rising (points 6-5) from +213 to +1980 m altitude, then descending over section 5-4 to +1279 m, then rising again over section 4-3 to +2010 m. The pipeline is laid underground at a depth of 75-90 cm. The narrow right-of-way was restored after the pipe was laid so that the land can be used.

The hydrotransport system is equipped with four pumping plants, with 13 piston pumps with rated powers of 1300 kW installed in pairs (one station has four, the others have three each).

In the first section the pump creates a pressure of 11.3 MPa, in the remaining sections 7 MPa, allowing the pumping plants to be spaced at distances of 96 to 128 km. The hydrotransport system carries 660 tons of coal per hour, and the pipeline is designed to operate around the clock.

The actual operating time factor of the pumps, which have valves made of wear-resistant alloys, is 99%, as compared to the planned level of 95%. The pumps transfer the slurry through the pipe at 1.7 m/s which, in the opinion of the American specialists, is the economically most favorable speed for the conditions.

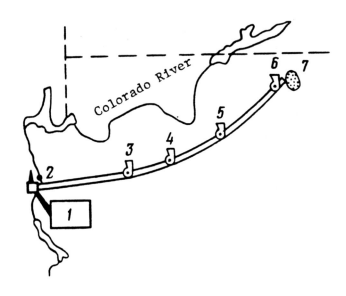

Figure 3.3. Diagram of Black Mesa coal pipeline: 1--electric power plant; 2--dam; 3-6--pumping plants; 7--coal mine.

Makeup water equal to 15% of the slurry flowing through the pipe is provided to the power plant system from the Colorado River.

At the Mojave electric power plant the slurry is dumped into reservoirs with a capacity of 22,700 m^3, where it is agitated by 370 kW agitators.

The system used to dewater the slurry in centrifuges is illustrated in figure 3.4. After dewatering in 40 centrifuges to a moisture content of 15% the coal is fed by belt conveyor to a mixer in which it is finally dried and then sent to the boilers of the electric power plant.

Since the system was put in operation (1970), no wear of the Black Mesa pipeline caused by the solid particles of material has been observed at all, indicating that ordinary steel pipe can be used for the hydrotransport of coal.

In addition to its great reliability and economy, the Black Mesa coal pipeline has other advantages unique to pipeline hydrotransport: there are no transportation fuel losses; there is no environmental pollution; the right of way can be used.

Four large pipeline systems with throughputs of 25 million tons per

53

year, extending over 1500 km, are to be constructed in the USA by 1985 for the transportation of coal from various regions in the nation to points of consumption. The transportation of this quantity of coal by rail would require an additional 150,000 railroad cars and 80,000 locomotives.

Other pipelines are also planned for the USA and Canada, information on which is presented in table 3.1.

Country	Diameter, mm	Length, km	Throughput, 10^6 t/yr
USA	950	1860	25
USA	750	1300	16
USA	600	290	10
USA	550	1200	9
Canada	600	800	12

3.2. Requirements Placed on Slurries

Long range hydrotransport is effective only if the coal particle size, concentration of slurry and conditions of its movement, as well as the equipment, are properly selected. Coal subjected to hydrotransport over long distances can then be used by direct combustion of water-coal mixtures in the boilers of electric power plants or as dry material after dewatering and drying in special installations. Depending on the methods and utilization of the coal, various particle size distributions must be used.

Slurries of high concentration, made of small diameter coal particles and intended for direct combustion, must meet certain requirements determined by the technological specifics of long range coal transport. The combustion of the coal must produce the required quantity of heat; the slurries must be stable, i.e., the coal should not settle out quickly if the hydrotransport system shuts down. If the slurry does separate into layers, the sediment must be easily picked up by

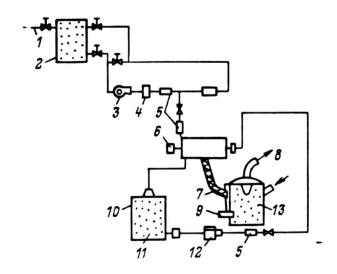

Figure 3.4. Slurry dewatering system of Black Mesa coal pipeline:
1--main pipeline; 2--collecting reservoir; 3--slurry transfer pump;
4--slurry densitometer; 5--electromagnetic flow rate meters; 6--centrifuges;
7--belt conveyor; 8--coal powder fuel; 9--air intake; 10--clarified
water; 11--settling tanks; 12--pumps; 13--mixer.

the water when the pumps are restarted; the slurry should be sufficiently
fluid that it can be transferred by centrifugal or piston pumps.

The combustion of water-coal mixtures in the boilers of electric
power plants does not require expensive dewatering and drying installations, but does require the creation of special high throughput burners
and other equipment. When water-coal mixtures are burned without preliminary dewatering and drying, a comparatively simple automatic continuous process can be achieved. The replacement of standard combustion
chambers with combustion of coal in layers by flame burners results in
losses due to incomplete burning of the coal (in layer burners the losses
amount to 9-14%, in flame burners up to 5%). Direct combustion of water-
coal slurries, both finely and coarsely dispersed, is possible.

The cost of electric power for the preparation of coal powder depends
on the particle size and represents about 1.5-2.5% of the total electric

55

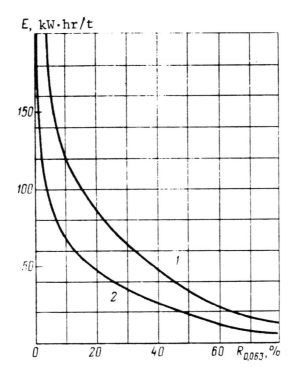

Figure 3.5. Specific energy consumption as a function of particle size of coal: 1--dry grinding; 2--wet grinding

power produced by the electric power plant. The power consumption for grinding depends on the particle size of the coal; the electric power consumption is least for coarsely dispersed slurries. However, increasing the coal particle size makes combustion of the coal more difficult and results in an increase in losses due to incomplete combustion in the boiler.

The influence of the fineness of grinding on the energy consumption of the process of grinding can be seen from figure 3.5, which presents the results of experiments performed at the A.A. Skochinskiy Mining Institute in a mill with steel grinding cylinders. The power consumption of fine grinding was studied using type I coal with an ash content of 13.2 and 17.8%. As the content of particles which remained on a screen with an aperture of 0.063 mm increased, regardless of the grinding process used, the energy consumption decreased rapidly, which is explained by the decrease in specific surface of the ground coal.

The minimum power consumption is obtained by the coarsest possible grinding of the coal; however, in this case, the fuel losses due to incomplete combustion increase. Maximum economy of the installation is achieved when the total of the cost of fuel preparation and combustion are minimized.

To determine the influence of particle size distribution of coal on the viscosity and stability of slurries, DonUGI Institute conducted special studies. We can see from figure 3.6 that as the particle size of a slurry increases, its viscosity decreases. With a content of the +0.063 mm fraction of over 25-30%, a slurry can be considered coarsely dispersed.

Figure 3.7 shows the results of experiments to determine the viscosity of coarsely dispersed water-coal slurries containing coal particles 0.06-0.2 mm in diameter. The coarsely dispersed slurries had lower viscosity than finely dispersed slurries. Finely dispersed slurries with high consistency act like viscous-plastic liquids: their viscosity depends not only on their physical and chemical properties, but also on the conditions of movement. The movement of these slurries involves high viscous resistance. Data on hydraulic resistance during movement of finely and coarsely dispersed slurries with a density of 1210 kg/m^3 are presented in table 3.2.

At low speeds, the hydraulic resistance of finely dispersed slurry is significantly greater than that of coarsely dispersed slurry, due to the greater values of structural viscosity. The difference increases with increasing concentration and decreases with increasing speed.

Long distance transportation should be performed with coarsely dispersed slurry, which involves lower power consumption for grinding of coal, as well as moving through the pipe with lower hydraulic resistance than is the case for finely dispersed slurry and assures normal operating conditions and high reliability of the hydrotransport system.

The layer separation time of a slurry depends on the physical and mechanical properties of the sediment formed. If the sediment forms a solid body, it is quite dangerous and threatens normal operation of the hydrotransport system. If the excess pressure created by the pump

upon startup of the system is
sufficient to cause the slurry
to begin movement, and the
sediment has not "caked", the
presence of a sediment represents no additional difficulty.

Detailed studies performed by
the Institute of Fossil Fuels
have shown that even with a
significant change in the initial
viscosity of the slurry (from
0.5 to 0.7), the moisture content
of the sediment formed varies
over relatively narrow limits.

Figure 3.8 shows the variation
and initial shear stress as a
function of density of a slurry

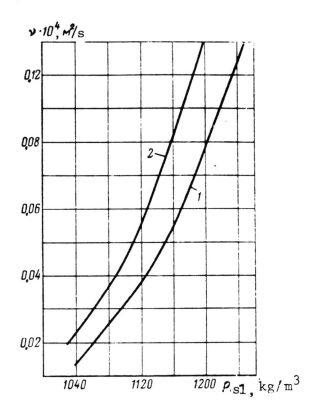

Figure 3.6. Variation in kinematic
viscosity of slurry as a function of
particle size distribution: 1--coarsely
dispersed; 2--finely dispersed

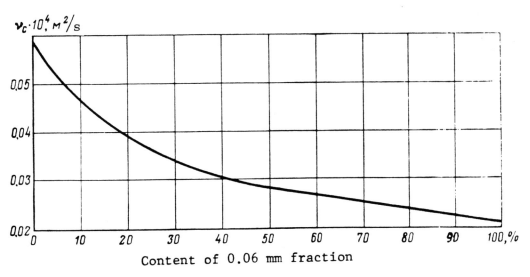

Figure 3.7. Kinematic viscosity of coarsely dispersed slurry.

58

(for more convenient comparison, θ_{st} is expressed in g/cm^2).

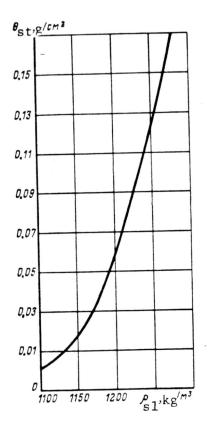

Figure 3.8. Initial shear stress as a function of slurry density.

In slurries which are intended for further hydrotransport, the initial shear stress may reach 0.11 g/cm^2; however, even after long-term shutdowns (several days) the slurry begins moving rapidly when the pump is restarted.

The sediment layer is a loose mass which is easily moved. The process of layer separation of coarsely dispersed slurries depends on the concentration of the slurry involved. As the concentration increases, the stability of the slurry increases. At a slurry density of 1170 kg/m^3, layer separation is practically complete after 100 minutes, whereas at 1240 kg/m^3 it continues for 160-170 minutes (figure 3.9), which is explained by the increase in the specific surface of the coal due to the larger number of solid particles.

As research has shown (figure 3.10), finely dispersed slurries have somewhat greater stability, i.e., separate into layers more slowly, than coarsely dispersed slurries; however, this advantage is not decisive, since coarsely dispersed slurries assure reliable operation of the hydrotransport system.

Coarsely dispersed slurries after long-term shutdowns (tens of hours) begin moving rapidly when the hydrotransport system is restarted. The sediment is broken up in 1-1.5 minutes, first carrying away those particles of the sediment located at the boundary with the liquid, after which the process of transportation of the sediment layer becomes increasingly intense until the entire slurry begins to move as a homogeneous fluid.

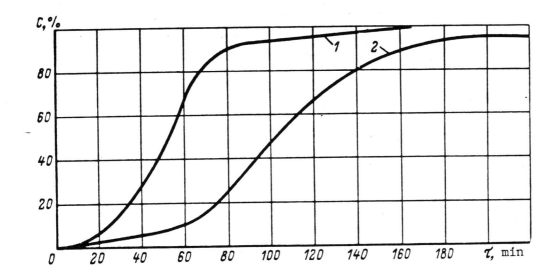

Figure 3.9. Influence of slurry density on layer separation time:.
1--ρ_{s1} = 1170 kg/m^3; 2--ρ_{s1} = 1250 kg/m^3

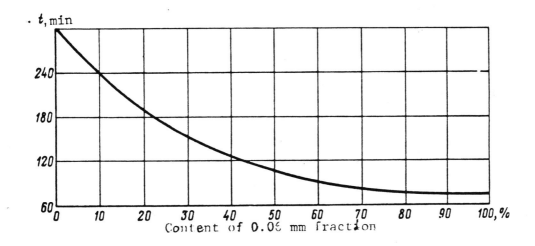

Figure 3.10 Influence of particle size distribution of slurry on
layer separation time

The ability to break up the structure rapidly means that when the hydrotransport system is started, very little excess pressure is required; therefore, coarsely dispersed slurries represent no particular difficulties in terms of startup conditions in comparison to homogeneous fluids. The weak structure of the sediment means that even short vertical sections, up to 2-3 m, do not hinder normal startup of the hydrotransport installation.

The tendency of coarsely dispersed slurries to separate into layers in hoppers can be utilized for thickening of slurries for direct combustion without preliminary dewatering and drying.

Table 3.2

Slurry	Speed, m/s					
	0.00	0.50	1.00	1.50	2.00	2.50
Finely dispersed	0.038	0.053	0.063	0.072	0.082	0.090
Coarsely dispersed	0.000	0.009	0.015	0.028	0.047	0.077

Slurries consisting of coal with particle diameters of 2 mm or less can be transported at densities of 1200 kg/m^3 or more. It has been experimentally established that the larger the particle size of the coal transported, the more dense the sediment formed when the hydrotransport system is shut down; for this reason, slurries of high concentration consisting of large-lump coal may cause plugging of pipelines upon transportation. The optimal concentration for transport of large-lump materials is usually not over 25% by volume.

When solid materials with a particle diameter of not over 0.1-0.2 mm are transported, the solid particles are usually uniformly distributed through the cross section of the pipe, and their maintenance in the suspended state requires very little expenditure of energy.

3.3. Preparation of Coal and Its Utilization at the Consumers Location

Selecting and designing equipment for grinding, one should first recall that ball mills are more productive; their effectiveness depends

on the grindability of the material, the particle size of the initial
and end products, the type and dimensions of the mill. Rod and ball
mills used for wet grinding have operating volumes of 0.9 to 85 m^3,
with electric motor powers of 20 to 2500 kW.

At thermal electric power plants, the powdered fuel is prepared
in ball mills of various types (ShBM, ShMG, etc.). Coal is ground in
a closed cycle, with the largest particles returned to the same mill
after the crushed coal has been classified. It has been found that,
regardless of the throughput of the beneficiation plant, the optimal
feed particle size for rod mills is 25-30 mm, for ball mills 5-8 mm.

The selection of a grinding system is based on the requirements for
the end product as a function of the various factors influencing them.

Equipment for crushing and grinding of particulate materials
should be selected considering the required particle size. For example,
if coal is to be used at thermal electric power plants, in which it is
utilized as powdered fuel, the coal should be ground by the supplier
to powder size, mixed with water and transported as a high concentration
slurry to the consumer.

If dewatering and drying of the coal are to be performed at the
power plant, there is no need to crush the coal to powder size before
transporting it.

Ball mills are produced with drum diameters of 0.9 to 7 m, lengths
of 0.9 to 2.3 m, operating volumes of 0.45 to 80 m^3. Ball mills for
wet grinding are produced with center unloading, drum diameter 0.9-4.5 m,
length 1.8-6.0 m, operating volume 0.9-85 m^3.

Wet grinding rod mills have the same drum dimensions and mill capacities
as ball mills.

Large fuel and power systems require the creation of special high
productivity coal grinding equipment (handling up to 500-1000 tons of
coal per hour), operating by wet grinding.

At present, coal is dewatered and dried with the same equipment
used at beneficiation plants; work is under way on the direct combustion
of water-coal slurries without preliminary dewatering or drying.

Drum-type direct-flow dryers have drums 1-3.5 m in diameter and 4 to 27 m in length. Dryers of this type have high throughput, are reliable in operation, their shortcoming being that large quantities of powder are carried away with the gases, up to 20% when flotation concentrates are dried. Dryers require large areas and capital investments.

Tube dryers are used to dry small noncaking materials. They have a shorter time of contact of the material with the gases than drum dryers, require smaller capital investment for construction of the buildings which contain them; their shortcomings include considerable loss of powder and high power consumption.

Direct action drum dryers fed with coal slime containing particles in the 2-0 mm fraction with an initial moisture content of 20-25% produce coal powder with a moisture content of 1.3-3%; when the particle size of the initial material is 1-0 mm and the initial moisture content is 20-26%, the final moisture content of the end product is 3-6%.

Calculations have shown that the drying of coal in drum dryers requires the consumption of about 3% of the coal fed to electric power plants.

Dewatering and drying of coal at beneficiation plants are expensive processes.

The "Kuznetskaya" central beneficiation plant, which processes coal from hydraulic mines, utilizes thermal drying of 0.13 mm concentrate (70% slime and flotation concentrate). The throughput of the drum dryers, which are 3.5 m in diameter and 27 m in length, is 120-158 t/hr as initial material, 22.5-24 t/hr as evaporated moisture when the moisture content of the material received for drying is 20-27.8%.

VUKhIN Institute has developed a technology for dewatering of coal slimes which consists in that the slime is sent directly to flotation as a dilute slurry before it is sent to jigging machines, after which the larger particles are separated from it. This system has reduced the content of solids in the recycled water by a factor of 2.5 and increased the throughput of vacuum filters.

63

In West Germany, all beneficiation plants dewater the finer fractions of coal in centrifuges with screw or vibration unloading of the sediment and throughputs of 8-400 t/hr. Steam heated vacuum filters are used to dewater unbeneficiated slime and flotation concentrate. Drying of coal is significantly more expensive than mechanical dewatering in centrifuges. If the cost of removal of water in a centrifuge under the conditions of a beneficiation plant is taken as 100%, the cost of the same process on a vacuum filter is 160%, in a thermal dryer -- 4700%.

In West Germany, unbeneficiated powder or fine concentrate is added to the moist beneficiated slime.

Problems of dewatering and drying of brown coal are of particular significance due to the increasing role of inexpensive types of fuel for fossil-fuel power plants in the eastern part of our nation, particularly in connection with the development of the Kansk-Achinsk fuel and power complex. The transportation of brown coal in its ordinary form represents significant difficulties. The brown coal used at the Sakhalin regional electric power plant, for example, has low calorific value (3710 kcal/kg, moisture content 15%, ash content 29.7%) and when the moisture content rises to 20-25% the coal completely loses its particulate properties. The transportation of this fuel, its unloading and processing are quite difficult. There is as yet no effective means to allow complete elimination of the difficulties and problems arising when brown coal is used at thermal electric power plants. The only means for the utilization of wet brown coal is to dry it in special drying installations. For example, when low-calorie lignites are used at power plants in Bulgaria, they must first be dried in special drying installations.

As the scale of coal-fired power production increases and the use of slime and wet intermediate products at these plants expands, it has been suggested that the fuel first be dried in drum-type gas dryers. If preliminary drying does not achieve positive results, it becomes necessary to improve the quality of the fuel significantly, decreasing its moisture content (to not over 20%) and ash content (to 30-35%).

64

The particle size of the fuel must also be increased, utilizing the 50-0 mm fraction. The task becomes particularly complex in rainy and snowy weather.

However, these methods do not solve the problem of assuring reliable operation of a power plant using brown coal, since the problem of loading the coal into the dryer and other apparatus remains.

A new trend is the use of hydrotransport for brown, high ash and wet coal. The solid fuel is mixed with water, producing a slurry which is easily transported, and dry screening is replaced by wet screening.

Thus, the primary task in this technological chain of hardware is to assure effective operation of centrifuges for fuel which contains a large quantity of clay particles, which clog up the units and parts of the centrifuge and make it impossible for it to operate normally.

It is assumed that after dewatering of the undersize material from beneath the screen in the centrifuge, the concentrate from the centrifuge can be fed to the powder preparation system of the boiler, while the centrifugate is sent to the settling tanks for subsequent use.

The use of hydrotransport requires liquefaction of the fuel; the slurry loses its capacity to plug equipment and be deposited in it.

According to Ye.G. Volkovyskiy and A.G. Shuster [5], as the fuel is transferred on conveyors and in other transfer equipment, the mean quantity of fines formed between the screening and crushing equipment is not great, not over 2% of the oversize product. Deposition and plugging should not occur in this area.

Existing sedimentation centrifuges can deliver concentrate with a moisture content of 9-29%.

It is usually recommended that all methods of possible mechanical dewatering be used, with drying of the fuel employed only if these methods are not effective. Thus, the primary task is to create effective methods of mechanical dewatering.

Brown coal begins to cake at a moisture content of 23%, and is fully caked with a moisture content of 27% and an ash content of 40-50%.

Brown coal, it is thought, could easily be made loose again by
adding 10-15% water.

In US hydrotransport systems, all of the coal is centrifuged and
dried. It would be more expedient to dry not all of the coal, but
rather only the finer fractions; the coarse fractions, after screening,
would then be sent directly to the powder preparation section. It
would be desirable when using brown coal not to utilize additional
drying, since this is such an expensive operation. However, this could
be achieved only with highly effective equipment for mechanical dewatering of the coal. The creation of a centrifuge suitable for dewatering
of fuels containing large quantities of fine clay particles is a
great task in itself.

Considering the great difficulties involved in the transportation of
brown coal, particularly those related to possible increases in its
moisture content, it may be desirable to use hydrotransport from the
mine to the power plant with subsequent treatment of the coal to achieve
the conditions required by the power plant.

The use of hydrotransport would require the development of methods
for dewatering the coal. Dewatering of the coarse fractions represents
no particular difficulty, since screens will do the job. The finer
fractions of coal require more complex equipment for dewatering, and
mechanical methods alone can reduce the moisture content to no lower
than 12-15%. Only thermal drying can bring the moisture content down
to 7%. Tube-type dryers are most common, although they have not yet
achieved fuel moisture contents of less than 9% and their cost is quite
high. Drum-type dryers also have significant shortcomings, including
high capital investment requirements and significant operating costs.

A number of branches of industry, including chemistry, nonferrous
metallurgy and others, utilize drying of materials in a fluidized bed.
One advantage of fluidized-bed dryers is that the hardware required
is simple in design, having no rotating parts. Less material is carried
away with the gases in these dryers than in other types.

Fluidized-bed dryers are vertical pipes separated by a horizontal grate into two parts. The material to be dried is dumped into the upper part, and hot gases are fed into the lower part, which fluidize and dry the material. The dried material is discharged through apertures in the side.

The throughput of fluidized bed dryers according to foreign reports is as high as 540 t/hr.

The effectiveness of combustion of coal in power plant boilers depends on the type of coal, its content of volatiles and other parameters; therefore, the individual characteristics of the coal used prevent the creation of universal standardized combustion chambers. The effectiveness of the operation of coal-fired electric power plants can be significantly increased if the composition of the solid fuel supplied to the power plant boilers can be held constant. Direct utilization of coal as a slurry of a predetermined composition is quite promising. A water-coal slurry maintains good stability of properties of the coal (ash content, solid particle diameter, particle-size distribution). Slurries can be reliably transported through pipes like liquid fuels. They can be produced from coal beneficiation wastes or from run-of-mine hard coal with subsequent wet grinding of the coal.

67

CHAPTER 4. PARTICLE SIZE REDUCTION OF COAL IN THE PRIMARY ELEMENTS OF TECHNOLOGICAL SYSTEMS

4.1. Sources of Particle Size Reduction

The most important elements of technological systems in which particle size is reduced include: hydraulic mining operations, hydrotransport from the mine base to the coal pumping plant, preparation of coal for hydrotransport by coal pumps or other hydraulic hoist equipment, hydraulic hoisting of coal from the mine to the beneficiation plant on the surface or to the consumer.

Coal is transported by nonpressurized gravity flow through flumes or pipes from the face to the pressurized hydrotransport equipment.

The larger fractions of coal are separated on screens before crushing in some hydraulic mines. Crushing of the larger fractions to a particle size of not over 50 mm is necessary to allow the coal to pass through the pumps, and is performed by various types of crushers.

The coal is hoisted to the surface primarily by coal pumps, less frequently by airlifts or pipe-type feeders.

As the coal moves through flumes and pipes, passes through crushers, coal pumps and other equipment, it is crushed -- its mean particle diameter is reduced and the content of micron fractions is increased.

The amount of particle size reduction depends on the physical and mechanical properties of the coal and the technological system of hydraulic mining and hydrotransport used. The particle size refers to the mean diameter of a particle in a portion of coal, as calculated by the equation

$$d_m = \frac{\Sigma \gamma\, d}{\Sigma \gamma} ,$$

(4.1)

where γ is the yield of any fraction, %; d is the mean diameter of the fraction, mm.

The relative particle size reduction of coal by fractions

$$A = \frac{\Sigma \Delta \gamma}{\Sigma \gamma_{init}},$$
(4.2)

where $\Sigma \Delta_\gamma$ is the total difference in the yield of coal by fractions between the initial and final products, %; $\Sigma \Delta \gamma_{init}$ is the total yield of the initial product, %.

The degree of reduction in the particle size of coal during grinding i_{gr} can be described as follows:

$$i_{gr} = \frac{\Sigma \gamma d}{\Sigma \gamma_1 d_1},$$
(4.3)

where γ and γ_1 are the yields of the individual fractions and of the crushed product in terms of particle size, %; d and d_1 are the mean diameters of the fractions of the same products, mm.

In addition to the degree of particle size reduction, the particle size reduction index A is quite important; this index describes the relative reduction in the total content of a given fraction of coal in a given sample,

$$A = 100 - \frac{\Sigma i_{gr} \gamma_{end} \, 100}{\Sigma i_{gr} \gamma_{init}},$$
(4.4)

where γ_{end} is the yield of coal in the end product after crushing.

The maximum particle size is more significant for industrial processes than the entire particle size distribution of the coal.

The Rosin-Rammler equation is generally recognized to be the most significant of the mathematical descriptions of the particle size distribution of coal in the process of crushing.

The primary equation for the kinetics of crushing under actual conditions was suggested by S.Ye. Andreyev [6], V.V. Tovarev and

69

V.A. Perov to express the content of the coarse fraction as a function
of the duration of crushing.

During crushing of actual materials, for which their degree of hetero-
geneity must be considered, as well as the influence of changes in particle
size and crushability of the material in the process of crushing, the
equation, obtained from the equation for the kinetics of crushing, is

$$R_t = R_0 10^{-Kt^\beta},$$

(4.5)

where R_0 and R_t are the number of solid particles before and after
crushing; K is a parameter describing the change in relative crushing
rate in the process of crushing, the numerical value of which may be
greater or less than one, depending on the properties of the material
being crushed and the crushing conditions; β is a parameter describing
the change in relative crushing rate as a function of the physical and
mechanical properties of the material crushed and the crushing conditions.

The basic features of the crushing kinetics equation can be used
to study the process of crushing of coal during hydrotransport; this
equation allows determination of the content of any given particle
size fraction during crushing of coal in a pipe if the experimental
coefficients describing the coal are known. These include primarily
the absolute crushing rate, by which we mean the relationship of the
mass of the particles of the large fraction which have been crushed by
a certain moment in time to the duration of the process of crushing
up to that time.

The equation implicitly establishes the relationship between the
parameters of hydrotransport and the particle size distribution of
the coal. Use of the equation requires special studies to determine
the experimental coefficients which influence crushing: the particle size
of the coal, the rate of movement and density of the slurry, the
hydrotransport distance, and the specific characteristics of the pipeline --
number and quality of joints, etc.

Some of these parameters cannot be taken arbitrarily, since their
selection is determined primarily by the technical requirements for

assurance of reliable operation of the hydrotransport system, including, for example, the speed and concentration of the slurry. The other parameters are determined primarily by considerations of economy of operation of the installation (for example, particle size of the material transported).

Practical use of equation (4.5) requires that the degree of influence of various factors on crushing be determined, and that the most important ones be distinguished.

4.2. Particle Size Reduction of Coal Before the Central Hydraulic Hoist

As coal is separated from the seam, it is intensively crushed. The particle size reduction is great and is observed no matter how the coal is mined, increasing with increasing mechanization.

The particle size reduction which occurs upon extraction is greater than that of all other elements in the technological system of hydrotransport and hoisting of the coal in hydraulic mines. Hydraulic mining produces 20-47% fine fractions (0-6 mm) at the face, less than that of machine mining (16-56%).

Particle size reduction of coal during extraction depends on the strength of the coal, the number of cracks in the coal and many other factors. Some of these factors are not accessible to strict theoretical analysis.

In the operating hydraulic mines in this country, where various methods of hydraulic extraction of coal are used, the change in particle size distribution of coal at the face is as shown in figure 4.1, in which the upper curve corresponds to the minimum particle size reduction, the lower curve to the maximum particle size reduction of the coal. We can see from these data that even with the minimum particle size reduction the content of the larger fractions is not over 25-30% [7].

In nonpressurized hydrotransport, it is primarily the larger fractions of coal, +50 mm, which are further crushed, the 50-25 mm fraction being much less affected, while the 6-0 mm fraction is ground very little (figure 4.2).

71

The degree of crushing depends on the sharpness of turns, the height of drops, as well as their number. In the classification and crushing unit, particle size reduction is quite intensive. No less than 50% of the undersize product is further ground in crushers after classification in mines; if screens are not used, all of the coal is fed directly into the crushers.

The types of crushers used do not provide satisfactory effectiveness of crushing in terms of either upper or lower limits of crushed product particle size. Hammer crushers create large quantities of the finer fractions. A comparison of type DZSh and DMSh crushers shows that the degree of particle size reduction in the DZSh crusher is 2.11, in the DMSh crusher, 2.97.

Crushers form large quantities of the finest particles: the 1-0 mm fraction is increased by 15-17%, the 6-0 mm fraction by 6-27%, and the +100 mm fraction disappears almost completely. As concerns the medium fractions (25-13 and 13-6 mm), their content remains about the same after crushing as before.

Thus, crushers create large quantities of the finest fractions by grinding the coarsest fractions.

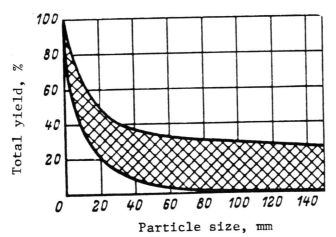

Figure 4.1. Change in particle size distribution of coal at the face.

Studies performed at hydraulic mines have shown that the greatest particle size reduction of transported coal occurs during classification and crushing. The total particle size reduction factor in this stage may be as great as 4.9, as opposed to 2.8 for all the other elements of the hydrotransport system combined, i.e., the particle size reduction occurring in classification and crushing may be as great as 65-70% of the total, the other 25-30% being distributed among all other elements of the hydrotransport system (taking the total degree of particle size reduction as 100%).

Nonmoving bar grates further reduce the particle size of coal. It has been found the mechanical classifiers and pick crushers are most effective in improving the grade of transported coal.

The great particle size reduction of coal in crushers and the formation of large quantities of finely ground particles lead to difficulties in further processing of the material on the surface.

Most of the particle size reduction of coal in hydraulic mines results from the fact that there is no effective equipment for preliminary classification of run-of-mine coal before crushing; therefore, over 50% of all coal mined is subjected to totally unjustified crushing and slime formation [7]. Mechanical classifying screens are needed to decrease the excessive particle size reduction of coal by crushing.

4.3. Particle Size Reduction of Coal in Pipes

Particle size reduction of coal was studied on a test installation with a four kilometer pipe (figure 4.3). Steel pipe 200 mm in diameter was laid out in 8 loops, each 500 m in length. Valves were installed to allow variation of the pipe length to 1, 2, 3 or 4 km.

The unclassified coal used in the study was fed into a hopper. A vibrating feeder dumped the coal onto a scraper conveyor which carried it to a container of known volume. Before the installation was started, the container and the entire system of pipes was filled with water, pumped in from an outside reservoir.

After the coal pump was started, an electromagnetic flow rate meter was used to determine the operating cycle length necessary to move a portion of coal throughout the entire hydrotransport system. The coal

was loaded into the system during one cycle, with the excess water
dumped into a collector.

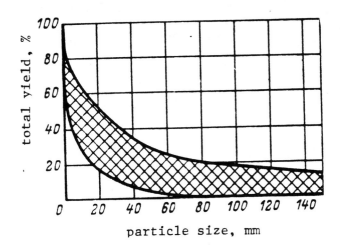

Figure 4.2. Change in particle size distribution of coal during
nonpressurized hydrotransport

Coal samples were taken with a drum-type sampler which sampled the
entire slurry flow. Several samples with a total volume of 2m³ were
taken throughout the entire loading cycle.

The speed at which the slurry moved through the pipe was maintained
constant throughout the study.

In addition to the electromagnetic flow rate meter, the flow rate
of slurry in the coal pipeline was also measured by a volumetric method
using a gauging tank and valve.

Samples of coal were taken after hydrotransport using the samplers.

The test installation included a sieve bend, from which the water
which had passed through the screen was sent to a slime water hopper,
then pumped to a vertical settling tank. The overflow from the settling
tank was sent to a volume-calibrated mixer. The dewatered coal was
screened, then sent to hoppers containing the various fractions, from
which it was carried by belt transporters out of the laboratory. The
water from beneath the screen was sent to a hopper and then on to a
waste water tank.

The specimens were screened on screens with apertures of 50, 25, 13,
6, 3, 1 and 0.5 mm. The 0.5-0 mm fraction was then further screened

into finer fractions: 0.3-0.5; 0.06-0.3; 0-0.06 mm.

The particle size distribution of the coal changed during the process of the studies, i.e. the yield of the individual fractions and the degree of crushing of each fraction changed, as did the overall degree of particle size reduction of the sample in the process of hydrotransport.

The particle size reduction of coal and pipes can be studied on a model of a coal pipeline consisting of a wheel test stand. The wheel test stand developed at UkrNIIGidrougol' Institute actually consists of three rings 3.2 m in diameter, made of pipe 100, 150 and 200 mm in diameter. The rings contain apertures which allow them to be filled with coal slurry. These apertures are covered with hatches which are fitted flush with the pipe. The wheels are rotated by a dc electric motor.

The electric motor is driven by a motor-generator installation, with a revolution counter on the free end of the shaft. The dc electric motor drive system allows experiments to be conducted over a broad range of speeds of rotation of the wheels of the test stand.

During the studies, the wheels are filled with slurry to two-thirds of their volume and rotated at a predetermined speed.

If the rotating speed is not too great, the walls of the pipe move at the rotating speed with respect to the slurry. It is thought that the solid particles perform movements quite similar in trajectory to actual conditions in a straight pipe.

The wheel stand does not completely simulate the movement of slurry in a straight pipe, since as the rings rotate, the movement of the slurry is observed in two layers, one adjacent to the walls of the pipe moving in one direction, while the core of the flow moves in the other direction; the solid particles in the ring also separate into layers depending on their particle size and density. Therefore, the concentration is different at different locations in the ring.

The differences in the nature of the movement of solid particles in ring pipes and straight pipes were studied by means of high speed cinematography in transparent models. In the wheelstand, the coal particles are more rapidly ground down than in a straight pipe, since

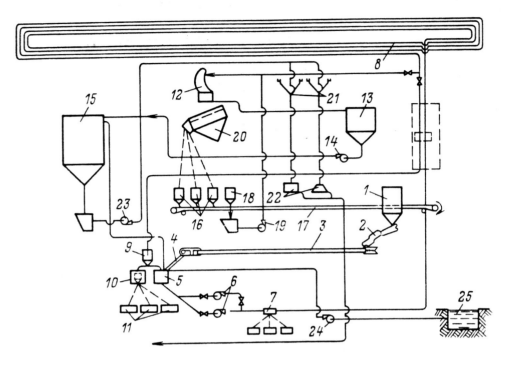

Figure 4.3. Test installation for determination of particle size
reduction of coal: 1--receiving hopper; 2--vibrating feeder; 3--conveyor;
4--chute; 5--calibrated container; 6--coal pump; 7--sampler; 8--pipes;
9--valve device; 10--gauging tank; 11--containers; 12-- sieve bend;
13--slime water hopper; 14,19,23, 24--pumps; 15--verticle settling tank:
16, 18--hoppers; 20--grate; 21--hopper; 22--feeders; 25--external
settling tank

one layer of coal moving upward rubs against the other layer, which
is moving downward. There are no oppositely directed flows of layers
of coal in a straight pipe; however, the influence of additional particle
size reduction caused by the joints in the pipe is quite great. These
factors balance each other to the point that in many experiments the
wheel test stand has yielded results quite similar to actual results
from the field. Experiments performed in pipes in straight lines are,
however, more reliable.

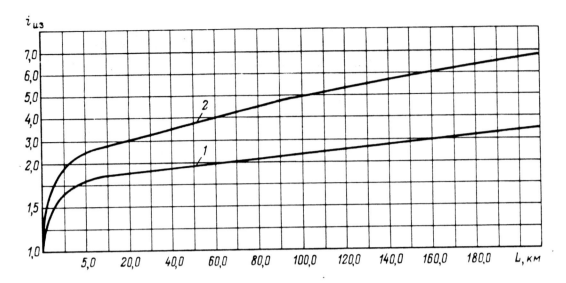

Figure 4.4. Influence of hydrotransport distance on degree of coal particle size reduction: 1--type "A" coal, 0-25 mm fraction; 2--type "G" coal, 0-50 mm fraction

4.4. Factors Influencing the Particle Size Reduction of Coal in Pipes

A great deal of experimental material has been accumulated on the particle size reduction of coal in pipes, indicating the existence of two characteristic zones as a function of transportation distance. In the attrition zone typical of short range hydrotransport, very rapid reduction in particle dimensions is observed, a result of the specific features exhibited by coal as a solid.

The surface of solid bodies is covered by large numbers of cracks and microscopic fissures; therefore, as the material is rolled the fracture resistance seems to increase.

Various forces determining the cause of particle size reduction in coal predominate, depending on the way in which the slurry moves. If the slurry moves in separate layers, particle size reduction is less, since the particles from the edge of one layer drop down and are picked up by the beginning of the next layer, friction of the particles against the wall of the pipe is not great and collisions between particles are not too frequent.

77

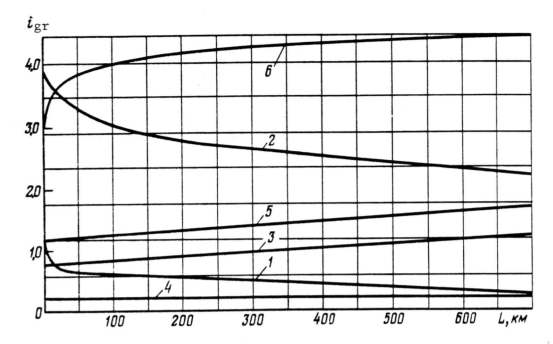

Figure 4.5. Variation in particle size distribution of coal in the 0-3 mm fraction as a function of hydrotransport distance: 1-- +3 mm; 2--1-3 mm; 3--0.5-1 mm; 4--0.3-0.5 mm; 5--0.06-0.3 mm; 6--0.06 mm

As the speed increases, the upper layer may move relative to the lower, nonmoving or slowly moving, layer. In this case, one layer rubs against the other. The particles in the upper layer are more intensively ground.

The nature of particle size reduction of coal during hydrotransport depends not only on the conditions of movement of the slurry, but also on the length of the pipe.

Figure 4.4 shows the variation in the degree of grinding of coal in the 0-25 and 0-50 mm particle size ranges as a function of hydro-transport distance.

Figure 4.5 shows the variation in particle size distribution of 0-3 mm diameter coal as it is transported through pipes over long distances. The 0.5-3 mm fraction is somewhat worn and its content in the initial section of the pipe decreases; other fractions are very little worn during hydrotransport.

Most intensive attrition occurs in short range hydrotransport (3-5 km) primarily due to cleavage of the material as a result of its many structural defects. The microscopic cracks and surface nonuniformities, etc. result in rapid particle size reduction as a result of splitting and "rolling down" of particles. Crushing of the material is characteristic of this zone, with very little wearing.

In the middle and long range hydrotransport zone (past 5 km), the significance of actual crushing decreases and processes of wear become more significant, until they represent the most important factor in the formation of micron-size particles.

It is difficult to establish a clear boundary between medium and long range hydrotransport, since the degree of particle size reduction depends on many factors which cannot be strictly considered.

The amount of particle size reduction of coal depends on the conditions of movement of the slurry, which determine the nature of the contact of the particles with each other and the walls of the pipe; the greater the number of particles which are in contact, the greater the particle size reduction. For this reason, movement at a speed higher than the critical speed, so that all of the particles are in the suspended state, causes more rapid particle size reduction than movement conditions in which a portion of the material remains on the bottom of the pipe as sediment.

Particle size reduction is greatly influenced by the properties of the materials, particularly their hardness and toughness. With otherwise equivalent conditions, as the hardness of particles increases they have less tendency to slip down out of the grinding zone and are therefore more easily broken; for this reason toughness is more important than hardness, since tough materials are less easily ground than hard materials.

The criterion of hardness is not very suitable for evaluation of the grindability of a material due to the significant role played by natural grains in grinding, since fracture occurs along natural grain boundaries. The smaller the coal particle diameter in comparison to the size of a natural grain, somewhere between 0.2 and 0.6 mm, the less likely the coal is to be further broken.

If the finely ground product is not screened out, the process of further particle size reduction is slowed down. As coal moves through a pipeline it is ground, and since the finely ground coal is not removed from the system, the degree of particle size reduction should decrease. This phenomenon also explains the decrease in the degree of particle size reduction of large lump coal when it is transported in slime.

The degree of particle size reduction of coal with an initial particle diameter of less than 0-6 mm is determined by the equation

$$i_{gr} = K_p \, \delta^X v \, c \, L,$$

(4.6)

where K_p is an experimental coefficient which depends on the properties of the slurry transported (dimensional unit, m^4/s); δ is the mean particle diameter, mm; v is the speed of movement of the slurry, m/s; χ is the exponent, which depends on the properties of the material transported; c is the concentration of the slurry by weight; L is the distance over which the coal is transported, km.

The degree of particle size reduction and the yield of coal in the micron fractions are most greatly influenced by the initial particle size of the coal and the distance over which it is transported: therefore, equation (4.6) can be represented as follows:

$$i_{gr} = K \, \delta^a \, L,$$

(4.7)

where K and a are experimental coefficients considering the properties of the slurry and the specifics of the hydrotransport installation.

The variation in the degree of particle size reduction as a function of pipeline diameter was tested using three ring pipes 100, 150 and 200 mm in diameter. It was found that the pipe diameter has practically no influence on the degree of particle size reduction; however, the influence of the pipeline on this process is significant, since joints and sharp turns in the pipe have a great influence on particle size reduction; it is difficult to consider all of these factors theoretically.

80

The findings indicate that in long range hydrotransport, particle size reduction depends primarily on the speed of movement of the slurry.

Hydrotransport of coal over long distances is performed at low speeds (1.6-2.2 m/s), allowing the energy costs, wear of pipes and equipment and degree of particle size reduction of the coal to be decreased. Usually the speed of movement of a slurry is the minimum speed which provides for reliable transportation, and its significance in particle size reduction is not decisive. The influence of the concentration of the slurry on the overall degree of particle size reduction has been established.

When concentation varies widely, it influences the particle size reduction of large lump coal, but not that of coal of the smaller fractions. It is possible that this results from the fact that large quantities of micron fractions are formed during hydrotransport, protecting the larger fractions from the grinding effects.

Experiments which have been performed have shown that the influence of slurry concentration on particle size reduction is manifested variously, depending on the particle size of the initial coal.

The degree of particle size reduction of the smaller fractions of coal in slurries of high concentration remains almost unchanged with increasing concentration. This is possibly explained by the fact that at high consistencies the freedom of movement of individual coal particles is severely limited, and the particles themselves have diameters no greater than the dimensions of natural grains.

The particle size reduction of coal depends on its physical and mechanical properties. The process, it is thought, is strongly influenced by the toughness, strength and jointing of the coal. There are no reliable methods for determination of the toughness of coal; the strength of coal considering the fine jointing, is determined by crush testing it. The Protod'yakonov strength yields satisfactory results in determination of the fracture resistance of coal, but fails to consider large grains separated by a distance greater than the maximum grain size.

Approximately 75% of the mined seams in the Donets basin with
lean and gassy coal, including coal intended for hydrotransport, as
well as anthracite, have strength ratings of 0.26 to 3.00. Stronger
coal should be ground less, though this has not been confirmed in
a number of experiments. For example, with otherwise equivalent
parameters the softer type "T" coal in the 13-0 mm fraction (f = 1.36)
is less ground (i_{gr} = 3.73) than type "A" anthracite of the same
particle size (f = 1.85; i_{gr} = 4.32). These properties are therefore
also not decisive in determining the degree of particle size reduction
of coal.

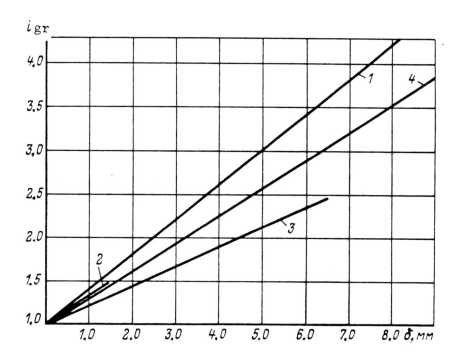

Figure 4.6. Influence of initial particle size on total particle size
reduction: 1--"A" coal, 0-50 mm fraction; 2--"D" coal, 0-2 mm fraction;
3--"T" coal, 0-50 mm fraction; 4--"G" coal, 0-50 mm fraction.

The presence of visible rock in the coal is quite important. The
properties of coal and rock influence particle size reduction differently.
Rock as it is transported may be transformed to the fine fractions if
it is wet or acts as a wearing body during transportation of coal (when

it has high strength).

The presence of rock in the mass causes the degree of particle size reduction of the coal to increase greatly. The particle size reduction of beneficiated coal during hydrotransport is much less. Whereas hydrotransport of beneficiated coal over a distance of 10-12 km results in an increase in the 0-0.063 mm fraction of 2%, hydrotransport of run-of-mine coal over the same distance causes the content of this fraction to increase by over 10%, i.e., by five times more.

Analysis has shown that particle size reduction is most greatly influenced by the initial particle size of the coal and the hydrotransport distance. Figure 4.6 shows the variation in the total degree of particle size reduction as a function of the initial particle size. For the same transportation distance, the greater the size of the initial coal particles, the greater the degree of size reduction. The increase in the content of micron fractions in coal which is transported is explained by the strong wearing effect of the larger fractions which, with otherwise equivalent conditions, are more rapidly broken down than the smaller fractions.

Studies which have been performed indicate that with an initial coal particle size of less than 3 mm, the increase in the fraction less than 60 μm in diameter is comparatively slight, greatly increasing with an increase in the initial particle size of the coal. As the mean particle size of the coal is increased by 1 mm, from 1 to 2 mm, the increase in the fraction less than 60 μm in diameter is 2%; this same increase in the 3-4 mm interval causes an increase in the smallest fraction of 11%.

The increase in micron fractions for the coal types studied can be expressed by the equation

$$\Delta_\gamma = 1.4 \, \delta_m^2. \tag{4.8}$$

Hydrotransport of the smaller fractions of coal (0-3, 0-2 or 0-1 mm) results in significantly less formation of the micron fractions than hydrotransport of large-lump coal. The increase in the content of 60 μm coal upon hydrotransport of 0-3 mm coal is not over 15%, whereas the increase during hydrotransport of 0-50 mm coal is over 70%. The greater degree of particle size reduction is explained by the fact that the large fractions break down more rapidly than the smaller particles; furthermore, they have a wearing effect on the smaller particles.

The total degree of particle size reduction depends on initial particle size as follows: 0-3 mm, reduction factor 1.75; 0-6 mm, 4.95; 0-50 mm, 13.1.

The influence of various parameters on the particle size reduction of coal less than 0-6 mm in diameter can be expressed by the following equations:

$$i_{gr} = n_1 \delta^{X_1};$$
(4.9)

transportation distance

$$i_{gr} = n_2 + m_2 L;$$
(4.10)

speed of movement of slurry

$$i_{gr} = n_3 + m_3 v;$$
(4.11)

concentration of slurry

$$i_{gr} = n_4 + m_4 C_p,$$
(4.12)

where $n_1 - n_4$, $m_2 - m_4$, and X_1 are experimental coefficients which depend on the properties of the coal and the characteristics of the hydrotransport system.

Slurries consisting of coal of not over 3 mm in diameter are transported over long distances by hydrotransport.

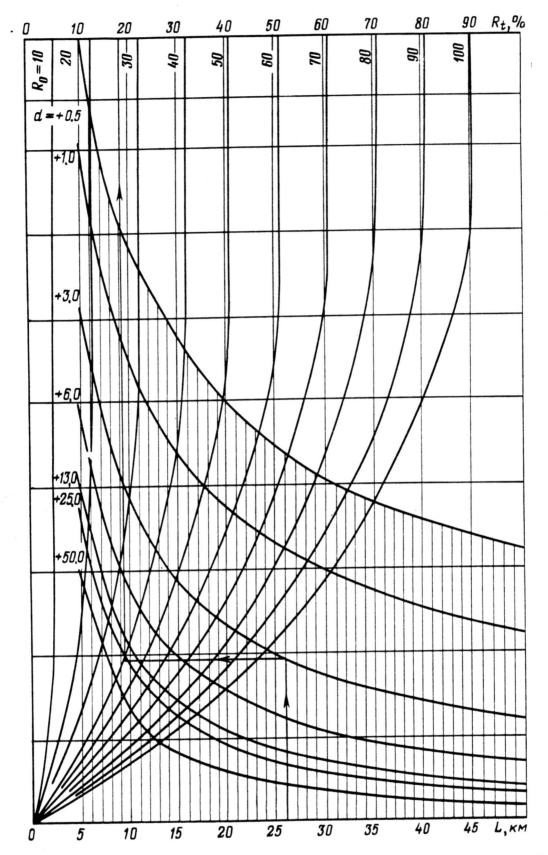

Figure 4.7. Nomogram for determination of particle size reduction of run-of-mine coal during hydrotransport (R_0, R_t represent the quantity of solid particles before and after grinding).

85

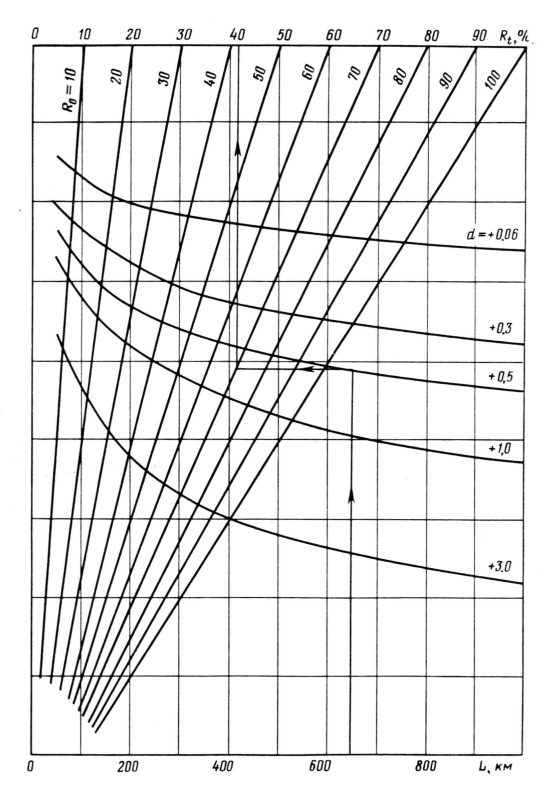

Figure 4.8. Nomogram for determination of particle size reduction of smaller fractions of coal during hydrotransport.

The process of attrition is significantly influenced by the structure of the materials and the shape of the particles; nevertheless, no relationship has yet been established between them.

Figure 4.7 shows a nomogram for determination of the particle size reduction of coal up to 50 mm in diameter during hydrotransport through pipes over distances of up to 50 km; figure 4.8 presents a nomogram for estimation of the particle size reduction of coal up to 3 mm in diameter for hydrotransport over distances of up to 1000 km. The nomograms were developed by candidate of technical sciences Yu.G. Svitlyy on the basis of experimental data obtained by UkrNIIGidrougol and equation (4.5).

4.5. Particle Size Reduction of Coal in Coal Pumps

Coal pumps are at present the most widely used means of hydrotransport. The process of particle size reduction in a centrifugal coal pump can be represented as follows: a coal particle is drawn by the stream of liquid into the impeller, and collides with the vanes. The speed of entry of the slurry into the impeller in some types of coal pumps is as great as 10-11 m/s; therefore, the damage to the coal by the entry edges may be significant.

Further grinding of the coal occurs in the channels and body of the pump. A coal pump has impact zones resulting from sudden changes in the direction of movement of the flow, including the point where the incoming flow encounters the vanes, rotation along with the impeller and the departure of the flow into the volute casing. The coal is also ground between the impeller and case, where the solid particles are crushed.

The influence of these factors on particle size reduction has been quite insufficiently studied, making the planning of coal pumps to minimize crushing quite difficult.

The degree of crushing of coal in a coal pump depends on the particle size and the concentration of the material transported. However, a coal pump cannot be considered a sort of centrifugal crusher, since the water does soften the blows of the particles of solid material and affect the trajectory of their movement. The crushing of coal in a coal pump depends not only on the speed of movement of the solid material at the intake, but also on the angular momentum of the slurry flow at the entry to the impeller part of the pump. The greater this momentum, the greater the particle size reduction.

UrkNIIGidrougol has studied the degree of particle size reduction of coal as it passes through a coal pump from 1 to 5 times, corresponding to the industrial conditions of hydraulic mines equipped with coal pumps connected in series.

The method of determination of the degree of particle size reduction of the coal in the coal pumps was as follows [8]. The coal pump operated under predetermined conditions corresponding to its nominal operating mode, obtained by creating counterpressure in the discharge pipe. The pressure was measured through a sampler. The test installation allowed the particle size reduction of the coal to be determined with the coal pump operating under various conditions.

The tests were performed as follows: the pump was turned on, the valve of the supply tank was opened to release the water, the valve of a smaller hopper was opened to release the coal into a larger container and the entire sample, after passing through the coal pump, was sent to a receiving container, where the coal precipitated and the clarified water was released into the large water tank of the laboratory by means of valves mounted in the side of the container. A millimeter grid was marked on the hatch of the receiving container in order to avoid loss of the smaller fractions of coal as the water was drained from the receiving tank. The mass of coal in each sample was 40 kg, the solid to liquid ratio 1:6 or 1:12. Particle size reduction of coal of three types was determined as it was passed through the pump one or more times. Figure 4.9 shows a graph of the particle size reduction of the coal after passing through the pump one to five times; the greatest particle size reduction occurs on the first pass through the pump.

After the first pass through the pump, with impeller diameter 460 mm and rotating speed 1450 rpm, the quantity of the largest fraction,

+25 mm, was decreased by a factor of 2.5, after a second pass -- by a factor of 4.6.

Particle size reduction is much greater for the larger fractions of coal, a result primarily of the breakup of lumps of coal into smaller lumps.

The intensity of particle size reduction of coal in a centrifugal coal pump depends on the mass of the lumps of coal, the speed of the flow at the input and output of the impeller. If the coal passes through several coal pumps, the total degree of crushing is expressed by the equation

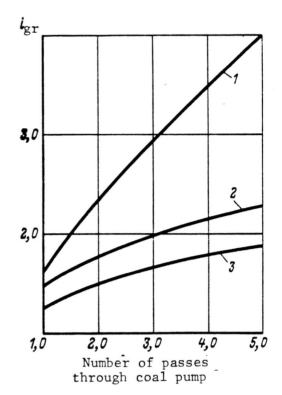

Figure 4.9. Particle size reduction of coal in centrifugal coal pump: 1--25-50 mm fraction; 2--13-25 mm fraction; 3--6-13 mm fraction

$$i_{gr} = 1.09 + 0.015 \, u_2 + 0.18 \, n_y.$$ (4.13)

where u_2 is the speed of the slurry at the exit from the impeller, m/s; n_y is the number of passes through the coal pump.

The crushing of the coal in coal pumps depends primarily on the number of times the coal passes through the coal pump in the process of hydrotransport. The influence of the remaining factors, the most important of which may be the concentration and the physical and chemical properties of the coal, should be studied in subsequent research.

A coal pump not only breaks the larger lumps, but also forms smaller fractions. The severe crushing of the coal during the first passes through a coal pump and the reduced intensity of crushing during subsequent passes are caused by the following factors: the sharp edges are broken off of the particles of coal and the weakest particles are fractured

89

first. The formation of fines creates a protective layer preventing fracturing of the larger particles. Crushing depends not only on the head created by the pump, but also on the design and the entry and exit speeds of the impeller.

In designing coal pumps, particular attention should be given to proper selection of parameters guaranteeing minimum particle size reduction.

Large quantities of fines are formed in a coal pump; the yield of the medium sized fractions, 25-13 and 13-6 mm ranges from 25 to 32% as the material passes through a coal pump up to five times.

We present here some data on the grinding of type "K" coal in the same coal pump with an impeller diameter of 510 mm and an operating speed of 960 rpm (table 4.2).

Table 4.2

Number of passes through coal pump	Initial fraction, mm					
	50-25	25-13	12-6	6-3	3-0	i_{gr}
1	39.8	20.3	9.3	8.1	22.5	1.84
2	23.0	20.0	10.0	9.2	37.8	2.60
3	11.3	19.5	11.3	11.3	46.6	3.60
4	8.2	19.8	14.4	11.0	46.6	3.88
5	6.0	17.0	16.5	12.0	48.5	4.55

The larger fractions are intensively ground in the coal pump, approximately half of the initial 50-25 mm coal being converted to the 3-0 mm fraction by five passes through a coal pump.

As the particle size decreases, the degree of grinding in the coal pump decreases. The degree of crushing depends on the physical and chemical properties of the coal, the number of passes through the coal pump and its parameters. As the impeller speed of a coal pump decreases (from 38.7 to 26.3 m/s), the degree of particle size reduction decreases.

At present, piston coal pumps are in use on operating long range hydrotransport installations; therefore, the study of particle size reduction in these machines is of particular interest.

A study was made of a type U8 piston pump, planned for the petroleum industry, which is more suitable in terms of its parameters for hydrotransport of coal over long distances than other available pumps. The U8 coal pump creates a pressure of up to 15 MPa, with a discharge of up to 162 m³/hr. Studies have shown that it operates reliably with coal particle diameters of up to 3 mm.

The method of determining the degree of particle size reduction in the piston pump was approximately the same as that used for the centrifugal pump. We present below data on the degree of particle size reduction of coal in the piston pump when slurries of coal of various particle sizes were fed through the pump (table 4.3).

The particle size reduction decreases with a decrease in the initial particle size (in the present case, the content of the 0.5-0 mm fraction by a factor of 1.5), the greatest particle size reduction occurring during the first pass, with the content of 0.5-0 mm fraction remaining practically unchanged in subsequent passes through the pump. This important fact means that there is a common picture of particle size reduction in centrifugal and piston pumps. The process of grinding of the finer fractions stabilizes due to the protection which they provide for the larger fractions.

Table 4.3

Entry material	6-0 mm fraction		3-0 mm fraction	
	d_m, mm	i_{gr}	d_m, mm	i_{gr}
Initial coal	2.09	1.00	1.94	1.00
After first pass	1.52	1.38	1.64	1.18
After second pass	1.38	1.51	1.44	1.35
After third pass	1.30	1.61	1.35	1.44
After fourth pass	–	–	1.27	1.53
After fifth pass	–	–	1.27	1.53

When finely ground coal with a particle size of up to 1-1.5 mm
is transported in slurries by either centrifugal or piston coal pumps,
the particle size reduction is slight and can be ignored.

4.6. Means of Decreasing Particle Size Reduction of Coal at
Hydraulic Mines

Both in the USSR and abroad, the technological systems used in hydraulic
mines are distinguished by the length and configuration of pipes, type
and quantity of equipment used. In the overwhelming majority of hydraulic
mines , the greatest part of the attrition of the coal in the process of
hydraulic mining and transportation occurs:

a) as the coal is separated from the mass (particle size reduction is
approximately the same for hydraulic mining and machine mining);

b) during gravity hydrotransport (primarily breaking down the large
+150 mm , 150-100, 100-50 mm fractions, with less reduction of the 50-25 mm
fraction). The yield of these fractions is decreased within hydraulic
mines by 4-20%, while the increase in the 6-0 mm fraction is up to 16%.

If the joints along the hydrotransport path are in unsatisfactory
condition, not only the large fractions, but also the medium fractions
of coal, 25-13 and 13-6 mm, will be ground down.

The most intensive process of attrition of coal is observed where
sections of the pressurized hydrotransport system are joined, where
there is a change in the direction of flow of the slurry. At these
locations, the large lumps of coal are reduced in size by impacts.
The higher the drop and the sharper the turns, the higher the degree
of particle size reduction.

Metal flumes with a slope of 0.05-0.08 produce the least particle
size reduction. Particle size reduction of coal during gravity hydrotransport can be reduced by decreasing the number of turns, improving
the quality of connections and decreasing the height of drops. Preliminary classification and crushing units are particularly important for
reducing the crushing of coal. Stationary bar grates installed before

crushers should be replaced with moving-bar grates. The grinding of coal in grates with continuously moving, self-cleaning grates is less than in nonmoving grates. The unsatisfactory operation of nonmoving grates is caused by the fact that their slots become plugged with rock, coal, wood, etc. Flat lumps of solid material drop through nonmoving bar grates into the coal pumps, resulting in stoppage.

Since the slots in the grates become plugged, up to 60% of the undersize product travels on with the oversize product to the crusher and is unnecessarily recrushed, leading to an additional increase in the 0-6 mm fraction by 20-25% over its initial level.

Moving bar grates GPK (grate, moving-bar) and PKO (feeder, classifier-dewatering) have been developed, in which the effectiveness of classification reaches 99%. The use of these grates decreases the overgrinding of coal in crushers and significantly decreases particle size reduction.

The most important trends for improvement in the quality of coal at operating hydraulic mines using hydrotransport are as follows:

improvement of the nonpressurized hydrotransport system by decreasing the number of turns over the entire length, creation of smooth connections, particularly the joints between flumes and slurry chutes and the flumes in accumulating roadways;

development of a more effective design for preliminary classifiers with high efficiency of screening and crushers which do not crush the coal too fine. Modern crushers should improve the grade of the coal by crushing of large fractions without further reducing the size of the smaller fractions.

If it is particularly important to preserve coal quality, coal pumps should be replaced with feeders.

A small crusher-classifier should be developed for hydraulic mines capable of both classifying and crushing the large fractions while observing rigid tolerances for the upper and lower limits of particle size of the crushed product, namely a content of the +75 (or +100) mm fraction in the crushed product of not over 1%, an increase in the yield of the 6-0 mm fraction of not over 5%.

CHAPTER 5. DESIGN OF PRESSURIZED HYDROTRANSPORT

5.1. Initial Prerequisites for the Selection of a Design Method

The method of design of pressurized hydrotransport includes the following main stages: determination of specific head losses and critical speeds during movement of coal slurries.

In selecting a method of design of pressurized hydrotransport, one must consider as a basis that at the present time the science of suspension-carrying flows cannot theoretically determine the losses of energy in a pipe and the critical speeds of movement of solid particles.

The development of hydrotransport techniques requires increases in the consistency of slurries. The special techniques and instruments required for such studies of highly concentrated (50-55%) slurries are not yet available. There is as yet but one method for calculation of the characteristics of hydrotransport - utilization of the materials from studies which are summarized and published as empirical or semiempirical equations.

These studies have allowed us to produce data for the design of hydrotransport systems under specific conditions; however, they have not been based on a special program intended for the development of methods for design of hydrotransport, suitable for the most commonly used operating conditions which are possible in practice in the coal industry.

Of the many practical methods for design of hydrotransport of coal slurries, preference is given to the method based on empirical formulas obtained in special experimental investigations. Since this method can be reliable only when large quantities of experimental data are available covering the most important possible conditions of movement of the slurry, the need has arisen for large numbers of full-scale experiments.

To summarize the experimental data, they must encompass the range of parameters encountered in industry (diameter of pipes, granulometric composition, consistency and speed of movement of the slurry).

Attempts to transfer the results of investigations from small installations to larger ones, due to imperfections in the theory of modeling, have generally not yielded the desired results; this has slowed down planning and construction of industrial hydrotransport installations.

The need has arisen for the creation of laboratory installations with full scale test stands, combining the primary advantages of industrial (reliability of data) and laboratory (determination of critical conditions, ability to stop slurry flow, study various parameters of hydrotransport, etc.) installations.

Full scale test stands are also used for the development of modeling criteria and to obtain initial materials necessary for conversion of the results of testing conducted with one set of pipes and slurries to others which may be used in industrial installations of varying size.

The method of experimental study of hydrotransport on full scale test stands using the theory of similarity does not, of course, yield equations for the characteristics of all forces participating in the suspension-carrying flow. The results produced are limited; nevertheless they can be of great significance for a deeper understanding of the essence of the interrelationships and regularities involved.

The use of the theory of similarity allows the results of experiments to be used directly to determine the significance of factors acting on the suspension-carrying flow; all factors cannot be considered, but rather only those which can be experimentally determined.

The theory of similarity reveals not only the qualitative, but also the quantitative aspects of the phenomena; it allows the most important factors influencing a suspension-carrying flow to be revealed, their interrelationships established and the flow model used to be refined and its basis improved.

In order to determine the hydraulic characteristics of suspension-carrying flows, the need has arisen for grouping of slurries similar in particle size distribution, for the conduct of a number of series of tests of characteristic groups of slurries, determination of particular similarity criteria for estimation of the influence of each of the important factors and the general similarity criteria encompassing the phenomenon as a whole.

In experimental studies and the processing of experimental data, an engineering method has been used for determination of the individual particular variations of critical speed and specific head loss as a function of a number of variables:

only the quantities necessary for calculation of the parameters of the hydrotransport installation (diameter of pipeline, speed of movement of slurry, specific head losses, maximum consistency of solid particles) are determined;

the accuracy of determination of the measured quantities is determined by the accuracy of the instruments used;

the calculation utilizes only those quantities obtained as a result of utilization of industrial measurement instruments;

since under industrial conditions fluctuations are observed in the particle size distribution, density of solid particles, speed of movement of slurries, etc., the calculation must consider the reliability factor; therefore we must deal not with local characteristics, but rather with integral characteristics, averaged over time and space;

the engineering method of determination of hydraulic parameters should provide proper calculation of the economic parameters, and it should be considered that the error in determination of hydraulic parameters does not lead to a change in economic values. Thus, even with an error in determination of specific head losses by 20-25%, the error in calculated specific cost of hydrotransport will not exceed 5-6%.

The method of engineering calculations should be developed for each type of slurry individually. Each slurry has its own characteristic properties; therefore, universal equations suitable for all types of slurries cannot be given. The long range hydrotransport department of UkrNIIGidrougol' Institute has undertaken a research program encompassing most possible conditions of application of hydrotransport in the coal industry. This program has involved studies of hydrotransport for coal of various particle sizes from +50 mm to finely dispersed powder less than 0.2 mm. Studies have been performed on various types of coal and rock with densities of 1.3 to 3.5 t/m^3 and various particle size distributions for hydrotransport in pipes of 100 to 600 mm diameter with concentrations   up to 0.50-0.52.

The method suggested is based on experimental materials produced on full scale installations, i.e., full sized pipes and natural coal. There is therefore no need to convert the experimental results from one set of conditions to another. Since the method is based on experimental data encompassing the entire range of industrial diameters, consistencies and speeds of movement of slurries and the coal most probable to be used in hydrotransport, it is quite reliable.

5.2. Experimental Determination of Particular Similarity Criteria

The method used has set forth the task of creating an experimental base with full scale test stands featuring: pipes of full scale diameter and great length; the possibility of testing and investigation of industrial equipment; the presence of coal preparation and processing equipment; the use of modern testing and measurement instruments.

The experimental hydrotransport stations of UkrNIIGidrougol' used for the study of the hydraulic parameters and technology of hydrotransport equipment is equipped with pipes 1-4 km in length, a substation with a dc generator, a measuring gallery, test stands for the testing of pumps and other equipment. The four-story building contains the following test stands and equipment:

a machine room (with test stands for studying particle size reduction of coal during hydrotransport by coal pumps and the transmission of water-coal mixtures through a pipeline);

a measuring gallery (with pipelines 100, 125, 150, 200, 250, 300, 350 and 500 mm in diameter, each 120 m in length, equipped with various testing and measurement instruments for the investigation of the hydraulic parameters of movement of slurries, comparative testing of instruments and of the local characteristics of slurry flows);

a four kilometer pipeline (external stand) designed for investigation of the particle size reduction of coal as a function of transportation distance and the hydraulic parameters of suspension-carrying flows;

a test stand with a 20 Gr-8 pump (power rating 1800 kW, pipes 400 and 600 mm in diameter) in a separate building (Figure 5.1).

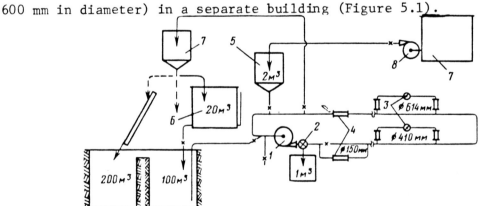

Figure 5.1. Installation for the study of head losses in a pipeline.
1 - coal pump; 2 - 3-way valve; 3 - instruments for measurement of head loss; 4 - instruments for measurement of flow; 5 - loading tank ; 6 - volumetric tank ; 7 - supplementary tank ; 8 - pump.

The electric machine room is intended for power supply of the experimental hydrotransport station units. The maximum power capacity is 3000 kW ac, 850 kW dc.

The measurement panel has two volumetric measuring tanks of 30 m^3 each, equipped with switching devices with automatic connection of instruments for determination of the flow rate. The experimental station is equipped with containers, centrifuges, vertical settling tanks, hoppers for unclassified coal and slime water. Glass pipe sections are available for observation of the process of movement of solid particles, with motion picture attachments.

The experimental hydrotransport station has the following advantages:

ability to study the equipment and parameters of hydrotransport in pipes of diameters most commonly used and planned in the coal industries;

investigation of equipment (coal pumps, water pumps, pipes) on test stands, on which the speed of movement of the slurry can be adjusted by changing the rotating speed of the electric motor shafts. This best method of adjustment is provided by the installation of dc electric motors;

the possibility of investigating hydrotransport with classified material, i.e., with predetermined particle size fractions of coal;

the possibility of studying particle size reduction of coal in pipes on two installations - the 4 kilometer, 200 mm diameter pipe and the wheel test stand loop with a loop diameter of 3.2 m and pipe diameters of 200, 150 and 100 mm;

the possibility of testing equipment (pumps, coal pumps) to determine hydraulic characteristics with operation on highly concentrated slurries;

the possibility of calibrating flow rate meters, consistometers, manometers and other instruments for hydrotransport;

in addition to full scale testing, studies can be performed on small glass models using cinematography to study the process of movement of suspended particles in pipelines;

the high capacity of the experimental hydrotransport station, sophisticated apparatus for measurement and recording of various quantities characteristic of the operating conditions, and the variability of the parameters of the installation allow the most important tasks in the area of hydrotransport of coal and other materials to be performed on this installation.

The specific head losses upon movement of water-coal mixtures depend on a number of parameters. Many experiments have produced data which relate to the flow of suspension-carrying streams with solid particles of various diameters, densities and particle size distributions in pipes with various wall roughnesses. Equations must be derived reflecting the results of these studies. To do this, special studies have been undertaken allowing determination of the influence of any one single factor on head loss and critical velocity, for example, slurry density, speed, etc.

These similarity criteria, which we shall call "particular" for head losses, are expressed as follows. For a model:

$$i'_v = \varphi_1(v) \; ; \; i'_{\rho_{CM}} = \varphi_2(\rho_{CM}) \; ; \quad i'_t = \varphi_3(t):$$

$$i'_{\rho_s} = \varphi_4(\rho_s) ; i'_a = \varphi_5(a);$$

$$i'_\beta = \varphi_6(\beta); i'_D = \varphi_7(D).$$

And for full scale equipment:

$$i''_v = \psi_1(v) \; ; \; i''_{\rho_{CM}} = \psi_2(\rho_{CM}); i''_t = \psi_3(t) \, ,$$

$$i''_{\rho_s} = \psi_4(\rho_s) ; i''_a = \psi_5 a;$$

$$i''_\beta = \psi_6(\beta); i''_D = \psi_7(D).$$

Within a certain range of the quantitites studied we can write:

$$i_v = \text{const}; i_{\rho_{CM}} = \text{const}; i_{\rho_s} = \text{const};$$

$$i_a = \text{const}; i_\beta = \text{const}; i_D = \text{const}.$$

Similarly for the critical velocities

$$v_{\text{кр}} = f(\rho_{CM}, D, k_V),$$

where k_V is an experimental coefficient considering the particle size distribution of the material transported.

$$v_{\rho_{CM}} = \frac{v'_{\rho_{CM}}}{v''_{\rho_{CM}}} = \text{const}; \quad v_D = \frac{v'_D}{v''_D} = \text{const}; \quad v_{KV} = \frac{v'_V}{v''_{KV}} \quad \text{const},$$

where $v'_{\rho_{CM}}$, v'_D, v'_{KV} are the particular similarity criteria for the model; $v''_{\rho_{CM}}$, v''_D, v''_{KV} are the same for full scale equipment; v is the speed of movement of the slurry, ρ_{CM}, ρ_S are the densities of the slurry and of the solid material; t is the temperature of the slurry; v_{kp} is the critical speed of movement of the slurry; D is the diameter of the pipe; α, β are experimental coefficients considering the particle size distribution and mean particle size of the solid material.

As a result of investigation of the movement conditions of coal and coal-rock slurries for the pipe diameters, solid material particle sizes and densities outlined above, the following relationships have been defined.

The function $i''_V = \psi_1(v)$ can be expressed by a one-term, two-term or three-term equation, depending on the required accuracy (Table 5.1).

It follows from the data presented that a two-term equation is suitable with accuracy sufficient for practice for processing of experimental data.

Table 5.1

Calculation Equation	Mean Square Deviation, %	Maximum Deviation, %
$i''_V = 0.01454\, v^{1.162}$	± 4.40	9.68
$i''_V = 0.0201 + 0.00322\, v^2$	± 2.76	5.55
$i''_V = 0.01153 + 0.0178\, v + 0.0079\, v^2$	± 2.60	5.36

The influence of speed is expressed by a two-term equation

$$i_{CM} = m + nv^X,$$

(5.1)

where m, n, X are experimental coefficients.

The function $i''_{\rho_{CM}} = \psi_2(\rho_{CM})$. Graphically this equation is interpreted with an accuracy of $\pm 5\%$ by lines (Figure 5.2) corresponding to the equation

$$i''_{\rho_{CM}} = k_{CM}\rho_{CM},\qquad (5.2)$$

where k_{CM} is the proportionality coefficient.

The criterion $i''_{\rho_{CM}}$ indicates that in two similar flows the density of the slurry should be identical. This criterion, obtained experimentally, is similar to the criterion s= constant

The function $i''_t = \psi_3(t)$. For each degree of increase in the temperature, the specific head losses decrease in relative units by only 0.05%; therefore within a narrow range of change of temperature its influence can be ignored in engineering calculations.

The function $i''_{\rho_S} = \psi_4(\rho_S)$. Specific head losses changed by 10-12% with variation in the density of the material transported (Figure 5.3); therefore a correction must be introduced to the calculations for density.

The influence of density can be considered as follows with an accuracy of ± 3-4%:

$$i''_{\rho_S} = k'_s \sqrt[3]{\frac{\rho_S}{\rho_0}} = k_s,\qquad (5.3)$$

where k'_S is the proportionality coefficient.

The function $i''_D = \psi_5(D)$. The graphic variation of specific head losses as a function of pipe diameter is shown in Figure 5.4. The equation is

$$i_D = \frac{k_D}{D},\qquad (5.4)$$

where k_D is the proportionality factor.

The function $i''_\alpha = \psi_6(\alpha)$ and $i''_\beta = \psi_7(\beta)$. The question of the influence of particle diameter on head loss is very complex. Various results obtained by various researchers attempting to determine the influence of particle size of the material transported on head loss have resulted from the fact that the so-called "mean" particle diameter was studied, which in itself is insufficient to describe polydispersed materials. The mean solid particle diameter describes the material being transported only approximately.

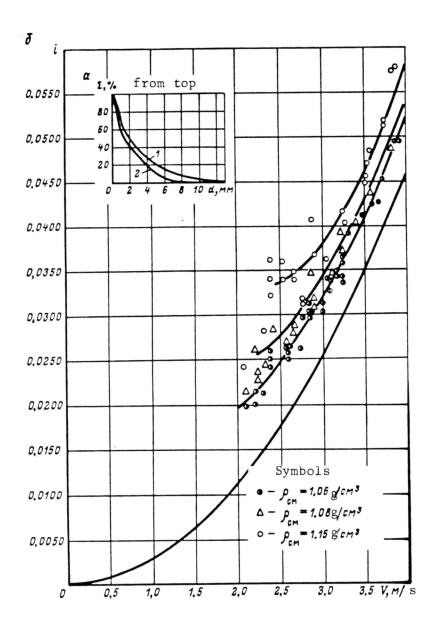

Figure 5.2. The influence of slurry density on head losses: a – screen analysis composition; 1 – before experiment; 2 – after experiment; b – function i=ϕ(v).

The influence of particle diameter on head loss is significant with a change of particle diameter of up to 2 mm. If the solid material consists of a mixture of very large and very small particles, it is possible to say what will happen, since the movement of the slurry depends to a great extent on the speed of its movement. With equal consistencies and speeds, head losses are determined largely by the particle size distribution of the materials transported. A change in mean particle size by a factor of 7 in our experiments decreased head loss by a factor of only 1.25 (Figure 5.5).

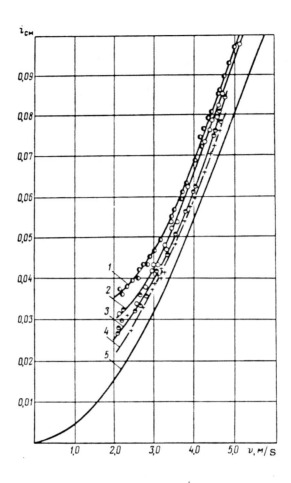

Figure 5.3. Influence of coal density on head loss: 1 – coal grade "A", ρ_S=1670 kg/m^3; 2 – coal grade "K", ρ_S=1360 kg/m^3; 3 – coal grade "G", ρ_S=1350 kg/m^3; 4 – coal grade "T", ρ_S=1300 kg/m^3; 5 – water.

When the slurry contains up to 14–16% of the larger fraction (+13, +50), up to 30% fine coal (<1 mm), a decrease in particle size by 2 mm causes a decrease in head loss by 6%.

Head loss is influenced by the content of large and micron size fractions in the slurry. Studies on slurries consisting of unbroken large lump material (rubber cubes measuring 17 mm on a side) and coal in the micron fraction have shown that in spite of the great density of the coal particles (1670 kg/m^3) in comparison to the rubber cubes (1180 kg/m^3) as the mean density of the solid material in the slurry increases due to an increase in the number of coal particles the head loss decreases due to a change in the content of the larger fraction.

103

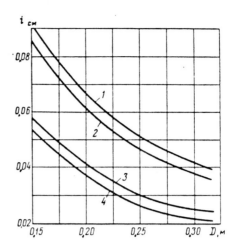

Figure 5.4. Influence of pipeline diameter on head loss: 1 - ρ_{CM}=1150 kg/m^3, v=4 m/s; 2 - ρ_{CM}=1080 kg/m^3, v=4 m/s; 3 - ρ_{CM}=1150 kg/m^3, v=3 m/s; 4 - ρ_{CM}=1080 kg/m^3, v=3 m/s.

The head loss is influenced by the micron fraction (88 μ) which is quite significant in the formation of the kinematic characteristic and the dynamic characteristics of the slurry flow. The larger fractions of solid particles do not have this great influence.

Solid particles of micron size form a homogeneous high density fluid with water. The carrier fluid in such a suspending flow is no longer water, but rather a finely dispersed slurry, with the larger particles present as a mechanical impurity. These slurries can be looked upon as consisting of a heavy carrier fluid, transporting large particles.

The heavy carrier fluid is formed by grinding of the material transported.

The content of micron fractions, within certain limits (up to 6%) has no influence on the head loss, then with an increase in the content of the fine fractions (>20%) the loss is stabilized.

The influence of roughness of the pipe walls is determined by performing experiments on hydraulically smooth pipes after preliminary polishing of the pipe walls by transmitting water-coal and water-sand slurries through the pipes for 100-150 hours. During this polishing time the resistance factor decreases by 20-25%, after which it remains virtually constant.

Studies of the conditions of movement of slurry consisting of coal of various particle sizes (maximum 100 mm) in pipes 100-600 mm in diameter have established:

the variation of specific head losses as a function of speed of movement of the slurry with mean square deviation ±2.76% can be expressed by a two-term equation;

the specific head losses are directly proportional to the slurry density;

the influence of the temperature factor on specific head losses can be ignored because it is so small;

the specific head losses are inversely proportional to the diameter
of the pipe;

the specific head loss is increased with an increase in density of
the material transported.

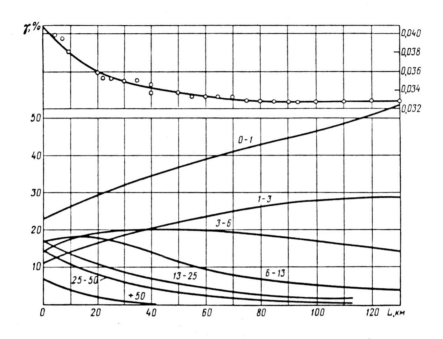

Figure 5.5. Influence of mean particle diameter of coal on head
losses.

5.3. Classification of Coal Slurries, Methods of Design of their Transportation

In the coal industry, coal of varying particle sizes and densities is
transported, meaning that water-coal slurries have various physical and
mechanical properties and follow various rules of motion.

Based on the general regularities of hydrotransport of water-coal slurries,
they must be subdivided into two main types - homogeneous and heterogeneous,
and the method of design of hydrotransport has been developed individually
for these two types.

Homogeneous slurries may be finely dispersed or coarsely dispersed.
Finely dispersed slurries consist of coal with particle diameters of not
over 0.2 mm with a content of the 0.063-0.2 mm fraction of not over 10%.

105

Coarsely dispersed slurries are those consisting of finely ground coal with a particle size of not over 0.2 mm with a content of 0.063-0.2 mm fraction of over 40%.

Homogeneous slurries also include water-coal slurries consisting of particles of 0-1, 0-3 or 0-6 mm with a content of the fraction −0.063 mm over 30%.

Heterogeneous water-coal slurries consist of coal in the 0-100, 0-70, 0-50, 0-25 or 0-13 mm fractions, as well as coal in the 0-6 and 0-3 mm fractions with a content of the fraction −0.063 mm of less than 30%.

Hydrotransport of heterogeneous slurries involves relatively high energy costs to overcome the hydraulic resistance of the pipes, as well as relatively high critical speed.

Hydrotransport of heterogeneous slurries at speeds less than the critical speed results in the formation of a nonmoving deposited layer of precipitated solid material in the pipe, which may lead to complete plugging of the pipe. Hydrotransport of heterogeneous slurries is accompanied by significant reduction in the particle size of the coal in the pipe. With a significant content of large fractions in the slurry (70-100 mm), the dimensions of which are comparable to the dimensions of the pipe, the danger also arises of plugging of pipes; therefore the diameter of the largest fractions of coal transported should be not over 1/3 the diameter of the pipe, and the consistency of heterogeneous slurries should not exceed 1:4-1:3 (S:L ratio by weight). The hydrotransport of heterogeneous slurries is primarily effective where distances are short.

Homogeneous slurries are analogous in the nature of their movement to homogeneous fluids of high viscosity and density; they can be characterized by the absence of a critical speed. The hydrotransport of slurries is desirable at high consistencies (with S:L ratio 1:1 by weight or higher) over distances of tens or hundreds of kilometers.

Slurries consisting of the fractions 0-1, 0-3 and 0-6 mm with a content of the 0-0.063 mm fraction over 30% are similar to homogeneous proportionally dispersed slurries, although they can be characterized by relatively smaller values of critical velocity (as for heterogeneous slurries).

5.3.1. Heterogeneous Slurries

Slurries consisting of materials obtained in mining or materials which have not undergone a special stage of preparation before loading into the hydrotransport system are heterogeneous.

One feature of hydrotransport is the variable conditions of operation caused by changes in the particle size distribution of the transported material. In the process of operation the slurries transported shift from one particle size group to another, determined by the transportation distance.

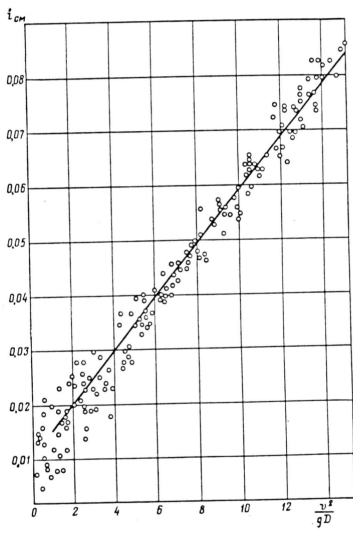

Figure 5.6. Variation and specific head losses as a function of Froude number

The head loss during movement of coal slurries is influenced by a number of factors, the degree of influence of each of which is different. Strongly influencing factors include: the speed of movement of the slurry, its density, the diameter of the pipe, the density of the solid materials. Weakly influencing factors include: the particle shape and size, absolute roughness of the pipe and temperature.

The particular dimensionless similarity criteria for head loss of heterogeneous water-coal mixtures have been used to obtain a general criterial equation which is presented in dimensionless form. The criterion of dynamic similarity for heterogeneous coal slurries must consider mass forces, regardless of the movement conditions.

Froude number   (Figure 5.6) can be used to process experimental data

$$Fr = \frac{v^2}{g\, d_c},$$ (5.5)

where d_c is the weighted mean particle size of the transported material. The similarity criteria, considering the general dynamic similarity criteria for homogeneous slurries, can be used to obtain an equation to determine the specific head losses:

$$i_{CM} = \frac{(1 + as)}{2g D}\, k_s\, (\alpha + \beta\, v^2),$$ (5.6)

where v is the design speed of movement of the slurry; k_s is a coefficient which considers the density of the transported coal:

Density of coal, kg/m³....1200; 1300; 1400; 1500; 1600; 1700
Coefficient k_s...........0.89; 0.91; 0.94; 0.96; 0.97; 1.00

$$a = \frac{\rho_s - \rho_0}{\rho_0},$$ (5.7)

where α and β are experimental coefficients which depend on the weighted mean particle size of the transported material, determined from the following experimental equations:

$$a = 0,087 + 0,000087\ \delta_{cp}^2\,;$$ (5.8)

$$\beta = 0,007 + 0,00027\ \delta_{cp}$$ (5.9)

(δ_{cp} is the weighted mean particle size of the transported material, mm).

For approximate calculations, α and β can be taken from the following table.

Particle size, mm	Mean diameter, mm	α	β
0-50	9-11	0,1120	0,0120
0-25	6-8	0,1100	0,0107
0-13	3-5	0,1053	0,0103
0-6	1.5-2.5	0,0967	0,0096
0-3	<1	0,0610	0,0075

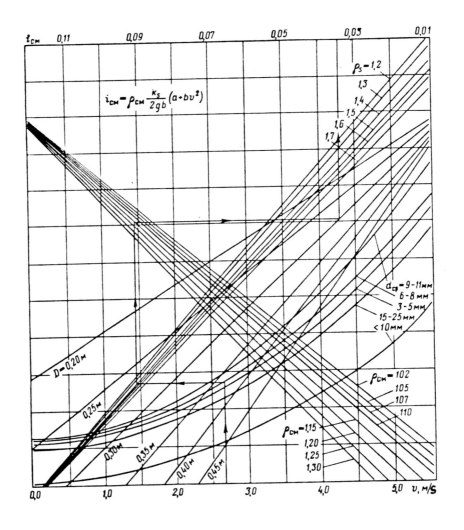

Figure 5.7. Nomogram for determination of head losses during hydro-transport of a coal slurry.

A nomogram (Figure 5.7) constructed from equation (5.8), is presented for convenience of calculation.

The head losses upon hydrotransport through vertical pipes are determined considering two components: losses to friction upon movement of the slurry through these pipes and the necessary pressure increment due to the weight of the column of slurry in the pipes.

The head losses are determined approximately by equation (5.8); there will be a certain reserve, since as the slurry moves through the vertical pipes the solid particles touch the walls of the pipes less than in horizontal pipes; however, this reserve will be expended in overcoming local resistances.

The weight of a column of slurry is determined by the equations:

$$i_\text{B} = i_\text{тр. B} + (\delta_\text{B} - 1); \tag{5.10}$$

$$i_\text{H} = i - (\delta_\text{B} - 1), \tag{5.11}$$

where i_B represents the full specific head loss in vertical pipes with rising slurry flow; i_H is the same for descending slurry flow; δ_B is the relative real specific weight of the slurry in the same pipes.

In sloping pipes, in contrast to vertical pipes, there is asymmetry in the distribution of velocities and consistency relative to the horizontal diameter. The head loss in slightly sloping pipes can be determined as in horizontal pipes while in steeply sloping pipes it must be determined as in vertical pipes.

When sloping pipes are used, one must consider that if the coal pump does not have sufficient excess head, the threat of plugging arises; to avoid this, these systems must be washed through with clear water before they are shut down.

The critical velocity in sloping pipes is less than in horizontal pipes; it should be calculated as

$$v_\text{cr.s} = v_\text{cr.h} \cos^2\alpha, \tag{5.12}$$

where $v_\text{cr.s}$ is the critical speed of the slurry in the sloping pipe.

For a vertical sector, the critical speed can be taken somewhat less, but this involves practical difficulties and is not actually desirable.

In the process of reaching the planned capacity of hydrotransport installations, slurry lines must frequently operate with sediment obstructing 60 to 70% of their diameter. Existing methods of designing pressurized hydrotransport of various bulk materials encompass conditions of operation with sediment obstructing not over 25% of the diameter.

The hydraulic resistance upon movement through a silted pipe depends on: the particle size distribution, particle size and density of the solid particles, the speed of movement, diameter of the pipe and depth of the layer of sediment. The layer of sediment does not remain stationary, but rather moves in the direction of movement of the slurry, and is continually renewed. If the diameter of the solid particles is not over 1–1.7 mm, the change in hydraulic resistance is determined by the degree of saturation of the slurry with solid particles alone and depends little upon their physical and mechanical characteristics or the diameter of the pipe. There is a certain height of the layer of sediment for which the hydraulic resistance is independent of the consistency of the slurry. If the depth of the layer of sediment is equal to or less than one tenth of the diameter, calculations can be performed as for an unsilted pipe.

The head loss is determined not only by the carrier fluid, but also by the concentration of solid particles, their mean density and particle diameter, as well as the number of micron diameter particles.

One feature of the equation is that it does not analyze the slurry as a "pure" carrier fluid with its own specific losses plus additional losses caused by the presence of solid particles in the slurry.

The convenience resulting from this largely artificial engineering approach to design is thus lost.

There is particular interest in determining the critical speed at which the hydrotransport fluid suspends the solid particles. The formation of plugs in the pipes upon hydrotransport should be eliminated, since this requires a shut down of the entire system, involves significant difficulties and requires the detection and elimination of emergency situations. The probability of plugging of a coal pipeline increases with an increase in consistency: the larger the fraction, the greater the density of the sediment and the more rapidly the process of compacting occurs. For coal with a particle size of less than 0.25 mm the density of the sediment is 1200–1300 kg/m^3 and the process of compacting lasts for up to 2 hours; for the 3–7 mm fraction the density of the sediment is 1300–1400 kg/m^3 and compacting is completed in 3 hours.

In horizontal pipes, plugs may form with slope angles of the pipe of over 5°. If a plug has already been formed, it can generally be washed down by a filtration flow or eliminated by the pressure created by the pump.

It has been established that continuous operation of the system is possible where the height of the deposited layer is not over half the diameter of the pipe.

The most important means for preventing plugs is correct laying of pipe. As the consistency of slurry increases, the vertical component of the speed of movement decreases in comparison with that needed to maintain fractions of a given sinking rate in the suspended state with the lower consistency.

Determination of the critical speed was performed by direct observation through small transparent inserts mounted in the pipe, with simultaneous recording of the critical operating mode using instruments to measure the flow rate of the slurry.

The studies established that in the zone of industrial use, the following dimensionless functions are valid:

$$v_{\text{кр.}} = \psi_1(s^{1/2}); \quad v_{\text{кр}} = \psi_2(D^{1/2}); \quad v_{\text{кр}} = \psi_3(\rho_s)^{1/3}$$

Processing of experimental data, the determination of particular and general similarity criteria have established the equation needed for determination of critical speed v_{cr} of a slurry consisting of coal material of varying grain size in horizontal or slightly sloping pipes:

$$v_{\text{кр}} = k_v \sqrt[3]{\frac{\rho_s}{\rho_0}} \sqrt{gD(1 + as)}, \text{ m/s}, \tag{5.13}$$

where ρ_s and ρ_0 are the density of the solid material and of the carrier fluid, kg/m^3; D is the diameter of the pipe, m; g is the acceleration of the force of gravity, m/s^2; s is the true volumetric consistency, as a decimal fraction; k_v is a coefficient considering the weighted mean particle size of the solid materials.

The following values of coefficient k_v can be used for approximate calculations:

112

Weighted mean particle size, mm   3-7   8-15   18

Coefficient k_v..............   1.15   1.27   1.44

For more precise calculations, the following equation is recommended

$$k_v = 1.12 + 0.0012 \; \delta_{cp}^2,$$

(5.14)

where δ_{cp} is the weighted mean particle size of the material transported, mm.

Equation (5.13) can be used with a weighted mean particle size of the material transported of over 3 mm, density of the solid material 1200-3500 kg/m^3 (coal and rock) and slurry consistency up to 20% by weight.

The operating speeds of slurry v are taken as 10% higher than v_{kp} with a weighted mean particle size of up to 8 mm, 20% higher with greater weighted mean particle size.

5.3.2. Homogeneous Slurries

The speed of movement of coarsely dispersed slurries is determined in consideration of technical and economic factors, based on the developed turbulent movement and minimum total (adjusted) capital and operational costs for the hydrotransport installation.

The minimum hydrotransport speed at which a water-coal suspension moves in the turbulent mode is determined by the equation

$$v_{min} = \frac{Re^* \, \nu_{CM}}{D}, \; m/s,$$

(5.15)

where Re* is the Reynolds number for slurries corresponding to the beginning of the developed turbulent flow (Re*=7500); ν_{cm} is the maximum possible kinematic viscosity of a water-coal slurry,

ρ_{cm}, kg/m^3.....	1320	1270	1200	1150
ν_{cm}, m^2/s	$0.39 \cdot 10^{-4}$	$0.11 \cdot 10^{-4}$	$0.08 \cdot 10^{-4}$	$0.06 \cdot 10^{-4}$

Hydraulic calculation of the parameters of transportation of coarsely dispersed slurries can be based on the usual formulas of hydraulics considering the density of the coal slurry.

The specific head losses are determined by the equation

$$i_{CM} = \lambda_{CM} \frac{v^2}{2gD},$$

(5.16)

where λ_{cm} is determined from the expression

$$\lambda_{CM} = \lambda_0' (1 + \beta'),$$ (5.17)

β' are the additional head losses caused by the presence of suspended solid particles in the flow.

The value of β' is influenced by certain hydrotransport parameters such as the density of the slurry, speed of movement, diameter of the pipe, particle size distribution, mean particle size and physical-chemical properties of the coal. Depending on density ρ_{cm} of the slurry, the values β' will be as follows:

ρ_{cm}, kg/m^3.................1320   1270   1200   1150

β'........................0.85   0.65   0.55   0.45

λ_0' is the resistance coefficient, used in the Nikuradze equation for new pipe

$$\lambda_0' = \frac{1}{(1.74 + 2 \lg \frac{D}{2\varepsilon})^2}$$ (5.18)

where ε is the absolute roughness of the inside surface of the pipe, m.

The specific head losses for homogeneous water-coal suspensions in horizontal, sloping and vertical pipes are the same.

Finely dispersed homogeneous slurries with high consistency have significantly greater kinematic viscosity than does water, and hydrotransport of these slurries is performed primarily in structured flows. The speed of movement of such slurries is approximately 1 m/s, regardless of pipe diameter.

Coal slurries of high consistency made up of finely dispersed coal do not follow the classical laws of friction, since their viscosity depends not only on the velocity gradient, but also on the structure of the slurries.

Usually for such slurries the Bingham-Shvedov equation

$$r_{CM} = \theta_{CT} + \eta \frac{dv}{dn} ,$$ (5.19)

can be used, where r_{cm} is the force of friction in the slurry; η is the viscosity coefficient; dv/dn is the velocity gradient.

Homogeneous coal slurries moving through pipes are characterized by the Reynolds number (Figure 5.8).

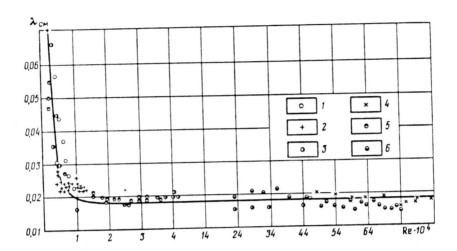

Figure 5.8. Variation in head loss as a function of Reynolds number: 1 - D=27 mm, ρ_{cm}=1260 kg/m^3; 2 - D=50 mm, ρ_{cm}=1300 kg/m^3; 3 - D=202 mm, ρ_{cm}=1240 kg/m^3; 4 - D=410 mm, ρ_{cm}=1080 kg/m^3; 5 - D=202 mm, ρ_{cm}=1240 kg/m^3; 6 - D=614 mm, ρ_{cm}=1080 kg/m^3.

For a structured flow, the Reynolds number according to the A. A. Skochinskiy Mining Institute, is determined from the equation

$$Re_{\text{стр}} = \frac{v\, D\, \rho_{\text{см}}}{\eta_{\text{стр}}\left(1 + \frac{\theta_d\, D}{8 v\, \eta_{\text{стр}}}\right)}, \qquad (5.20)$$

where $\eta_{\text{стр}}$ is the structural viscosity of the slurry assumed for approximate calculations to be 0.02-0.05 N·s/m^2 (for pipes 0.1-0.2 m in diameter); θ_d is the dynamic shear stress, determined with an RV-8 rotation viscosimeter.

The specific head losses upon movement of finely dispersed slurries are determined from equations (5.18) while the hydraulic resistance coefficient λ_{cm} is determined from the expression

$$\lambda_{\text{см}} = \frac{64}{Re_{\text{стр}}}. \qquad (5.21)$$

The use of the Reynolds number is possible if the coal slurry is looked upon as a homogeneous fluid with an effective viscosity which is higher than that of a carrier fluid.

Let us present a few characteristic graphs describing the flow of coarsely dispersed water-coal slurries in 614 mm diameter pipe (Figure 5.9). We can see from the graphs that with near zero speeds the specific head losses approach zero, with a slurry density of less than 1200 kg/m^3 the structural viscosity is either totally absent or the structure breaks down when exposed to slight excess pressure.

Coarsely dispersed coal slurries of low consistency practically do not differ from homogeneous fluids as they move through large diameter pipes; there is either no structural viscosity at all or, if present, it is broken down easily by mechanical forces, even at low transportation speeds [9].

At consistencies of over

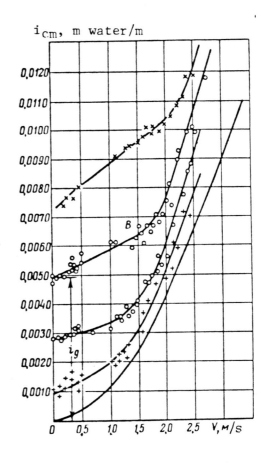

Figure 5.9. Specific head losses upon hydrotransport of a slurry in pipe 614 mm in diameter.

30% by weight, coarsely dispersed coal slurries move in large diameter pipes as anomalous, nonnewtonian fluids.

With a well-developed turbulent flow, the structure of a coal slurry breaks down and the initial shear stress approaches zero. In this case the viscosity becomes independent of the conditions of motion of the slurry, and is influenced only by the physical and mechanical properties.

Coarsely dispersed water-coal slurries, when static, separate into liquid and solid phases; nevertheless the slightest perturbation restores the homogeneity of the particle size distribution through the cross section of the pipe and movement can occur at very low speeds.

116

The speed of movement of a slurry in long distance coal pipelines can be approximately calculated at 5-7% above the critical speed.

The parameters of hydrotransport of coarse water-coal slurries should be selected considering the particle diameter of the coal, consistency and speed of the slurry and their joint influence on the technical and economic factors, since extremely fine grinding wastes energy.

Suspensions may flow in the laminar-structured mode with specific head losses greater than those in coarsely dispersed slurries.

It is also not practical to use slurries consisting of extremely large particles, since the energy costs increase in this case. These slurries are unstable not only in the static mode, but also in hydrotransport at low speeds.

The specific head losses involved in the movement of a water-coal slurry consisting of coal with a particle diameter of 0-3 mm with over 20% particles smaller than 0.063 mm present can be assumed to be identical, regardless of the hydrotransport distance.

Reliable operation of a hydrotransport installation requires that the speed of movement of the slurry be above the critical speed. The value of "reserve" required depends on the uniformity of feeding of the coal into the hydraulic system.

CHAPTER 6. METHODS OF HYDROTRANSPORT

6.1. Centrifugal Pumps

6.1.1. Parameters, Design Features

One of the most important types of equipment used in hydraulic mines is the centrifugal coal pump, used for hydrotransport of coal.

Coal pumps are designed for hydraulic hoisting of neutral coal slurries from shaft mines utilizing hydraulic mining, for hydraulic stowing and for hydrotransport of slurries to the surface with a normal temperature of the slurry transferred +20°C, permissible temperature limits +3 to +30°C.

The following types of coal pumps, the parameters of which are presented in table 6.1, are currently in series production at the Yasnogorsk Machine Plant.

Coal pumps designed for transportation of coal from shaft mines to the surface operate in conjunction with pipes of the following diameters.

Pipe diameter, mm	150-200-125	225-275-300	300-325-375	375-400-425
Capacity, m³/hr	350	600	900	1400

In contrast to other pumps designed for the transfer of suspended materials, coal pumps have high head at the impeller, as much as 175 m water. All coal pumps (except for the two-stage 12 UV6) are single-stage cantilever centrifugal pumps with horizontal shafts, widely used due to their great reliability and the accessibility of rapidly wearing parts.

Coal pumps have the following basic features, resulting from the need to transport large-lump materials: increased impeller width; small number of vanes on impeller (usually not over 4); increased radial gap between scroll and impeller. Furthermore, a number of design features are incorporated in coal pumps to decrease the hydroabrasive wear of pump parts. All of these features result in deterioration of the hydraulic characteristics in comparison to pumps designed for ordinary fluids.

118

Table 6.1

Characteristics	Coal pump							
	Single-impeller				Two-impeller			
	10U4	10U5	12U10	14U7	12UV6	12V6a	12UV66	14UV6
Capacity, m³/hr	350	600	900	1400	900	800	700	900
Head, m	120	175	83	175	320	280	250	320
Operating speed, rpm	1485	1485	1485	1485	1485	1485	1485	1485
Motor power, kW	320	630	320	1200	1600	1200	1000	1200
Impeller dia-meter, mm	620	690	510	715	700	660	630	650 and 7
Maximum particle size, mm	80	100	90	100	100	100	100	100
Maximum slurry consistency (by weight)	1:5	1:5	1:4	1:5	1:5	1:5	1:5	1:5

The presence of solids in the slurry also requires changes in the operating process of the coal pumps. Two basically different designs of high-head coal pumps are possible: a single-stage design with dual-flow or cantilever impeller and a two-stage design. Single-impeller systems with a cantilever impeller, with axial force relief, are most popular.

Coal pumps with dual-flow impeller have important advantages, achieving almost complete axial force relief, technological simplicity and low cost due to the symmetry of the shape of the body, ease of servicing and easy replacement of worn parts.

Figure 6.1 illustrates the most common design of a single-impeller cantilever coal pump. As studies by VNIIGidromash have shown, the optimal operating condition for a coal pump is determined by the hydraulic qualities of the volute casing. By changing the cross-sectional dimensions of the casing, one can significantly change the optimal flow rate.

Studies of the flow structure in a spiral-type of volute casing have shown that at flow rates below the optimal flow rate, circulating masses of liquid arise, making more than one revolution before entering

119

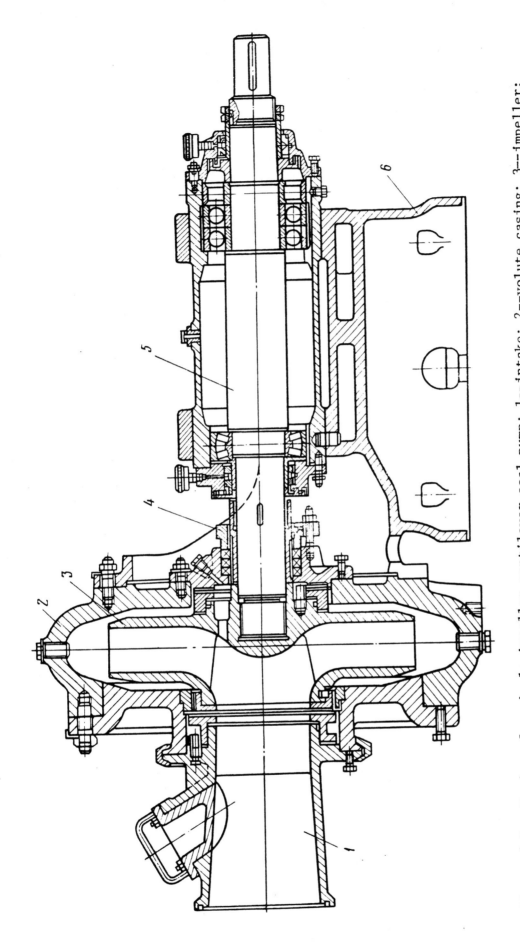

Figure 6.1. Design of a single-impeller cantilever coal pump: 1--intake; 2--volute casing; 3--impeller; 4--seal; 5--shaft; 6--frame

120

the diffuser; at flow rates higher than the optimal flow rate, a portion of the liquid flows into the diffuser, causing reverse flow near the tongue of the casing. Liquid from the impeller enters the casing non-uniformly around the perimeter, and the distribution of pressure on the impeller is also nonuniform.

The required reduction in the number of vanes on the impeller of pumps used for suspended substances reduces their efficiency and suction capacity. The efficiency of pumps with 3- and 2-vane impellers is 93.5 and 86% of the efficiency of a pump with a 6-vane impeller.

6.1.2. Regulation of Feed

Normal operation of hydraulic mines requires that the pumps used for water supply and the coal pumps used to return slurry to the surface be coordinated, which is achieved by changing their operating conditions by various methods of regulation. The operating conditions may be changed for brief or extended periods of time; the methods used for regulation vary depending on the duration of the proposed change.

Regulation of pumps is widely known and has been studied in considerable detail; the problems of regulating coal pumps have been less studied. There are some materials on the regulation of coal pumps by choking the discharge pipe, though this method has significant shortcomings; it is not economical, the installation of valves limits the maximum size of material to be transported, and the valves themselves wear quickly.

The characteristics of a coal pump for the transportation of coal slurries can be changed by changing the rotating speed of the pump shaft or by trimming the impeller.

As the rotating speed of the shaft is changed, assuming the efficiency remains the same, the parameters Q, H and N change as follows:

$$Q_2 = Q_1 \frac{n_2}{n_1}; \quad H_2 = H_1 \left(\frac{n_2}{n_1} \right)^2; \quad N_2 = N_1 \left(\frac{n_2}{n_1} \right)^3, \tag{6.1}$$

where Q_1, H_1 and N_1 are the discharge, head and power of the coal pump at shaft rotation speed n_1; Q_2, H_2 and N_2 are the same parameters for shaft rotation speed n_2.

If the impeller is trimmed, Q, H and N change as follows:

$$\frac{Q^*}{Q} = \left(\frac{D^*}{D}\right)^{1,6}; \quad \frac{H^*}{H} = \left(\frac{D^*}{D}\right)^{2,5}; \quad \frac{N^*}{N} = \left(\frac{D^*}{D}\right)^{4}, \tag{6.2}$$

where Q, H and N are the discharge, head and power of the coal pump with the standard impeller, without side diameter D; Q^*, H^* and N^* are the same parameters with the impeller trimmed to diameter D^*.

The method of changing the operating conditions of the coal pump by modification of the impeller by partial obstruction of channels, which can be used for long-term adjustment of coal pumps to nonstandard operating conditions, has a number of advantages over other possible methods. Studies have been performed to determine the parameters of coal pumps after partial blockage of the channels in the impeller to determine the extent to which the characteristics (discharge, head and power) change as a function of the clear path through the impeller, changed by partial blocking of channels [8].

This function, shown in figure 6.2, can be used in the design of impellers for coal pumps with smaller numbers of vanes. Plugging of channels leads to some deterioration in the hydraulic quality of the impellers and an increase in losses, expressed as a decrease in head at 0 throughput.

The ratio of the power and head with the valve closed (p_0 and H_0) to the power and head corresponding to the maximum efficiency (p_H and H_H) is as shown in table 6.2.

At 0 throughput, obviously, both the head and the power of the coal pump will decrease. Partial blocking of channels leads to some decrease in efficiency, which is indicated by the following experimental data.

122

Plugging of channels	No plugging	Plugging of		
		One	Two	Three
Efficiency	0.58	0.56	0.53	0.50

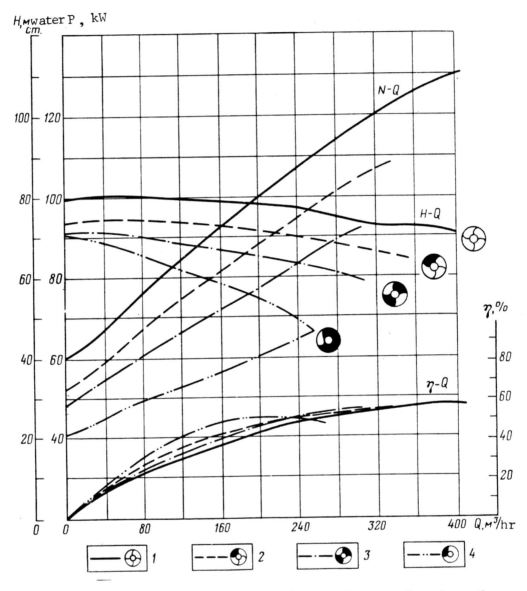

Figure 6.2. Change in flow through a coal pump due to plugging of a portion of the impeller: 1--no plugging; 2--plugging of one aperture; 3--two apertures; 4--three apertures

The studies determined that:

the decrease in cross section of the impeller due to partial plugging of channels with unchanged profile and a small number of blades leads to a decrease in throughput, head and efficiency of the pumps;

it can be considered for practical purposes that the decrease in throughput and head are such that the speed factor of the coal pump remains constant, corresponding to the optimal operating mode of the coal pump with the impeller not plugged;

plugging of a portion of the channels yields impellers with large open apertures.

The shortcomings of this method include the impossibility of smooth regulation of the flow while the pump is in operation.

Table 6.2

Number of channels plugged	Q, m^3/hr	Head, m			Power, kW		
		H_0	H_{nr}	$\frac{H_0}{H_H}100\%$	P_0	P_H	$\frac{P_0}{P_H}100\%$
None	380	79.5	72.5	110	58	130	44.5
One	320	73.0	67.5	108	51	107	47.5
Two	280	71.0	51.0	117	48	90	58.5
Three	220	70.0	52.5	133	40	63	63.5

Regulation by partial plugging of the channels of impellers can be used in hydraulic mines when it is necessary to maintain a reduced throughput for a long period of time or when new coal pumps of low capacity but large cross section of impeller apertures are planned.

6.1.3. Conversion of Characteristics

The manufacturers of pumps for suspended substances, including coal pumps, usually provide their characteristics for operation with water. Proper selection of these pumps to meet the conditions of actual operation require that their operating characteristics be converted from water to slurry.

The slurry head

$$H_{s1} = k_H H_0 (1 - as) \text{ м. water,} \qquad (6.3)$$

where k_H is a dimensionless coefficient which is independent of the discharge of the coal pump; H_0 is the head of the coal pump with water; s is the volumetric consistency of the slurry; a is [omitted from original Russian document! --Tr.]

For the types of centrifugal coal pumps presently used, coefficient k_H depends only on the consistency of the slurry and is determined by the equation

$$k_H = 1 - 0.6 \, s. \qquad (6.4)$$

Determination of the power N_{s1} consumed by the pump as it transfers slurry can be performed with the following equation if we know the power consumption of the pump for operation with water N_0:

$$N_{s1} = k_N N_0 (1 + as). \qquad (6.5)$$

Conversion of the efficiency of the pump for slurry η_{s1}, if the efficiency for water η_0 is known, can be performed as follows:

$$\eta_{s1} = k_\eta \, \eta_0, \qquad (6.6)$$

where $k_\eta = 1 - 0.2 \, s$.

The discharge of the pump Q_{s1} for slurry at identical head is decreased as the consistency increases in comparison with the discharge Q_0 of the same pump operating in water, which is considered by the discharge coefficient

$$Q_{s1} = k_Q Q_0. \qquad (6.8)$$

The discharge coefficient is determined by the equation

$$k_Q \, k_\gamma k_H = k_N \, k_\eta,$$ (6.9)

where k_γ is a coefficient which considers the change in density of the slurry

$$\rho_{sl} = k_\gamma \, \rho_0.$$ (6.10)

The values of the centrifugal pump characteristic conversion coefficients for coal slurry (coal particle size 0-3 mm) are presented below:

s = 0.05, 0.100, 0.150, 0.200, 0.250; 0.300, 0.350, 0.400, 0.450; 0.500

k_N = 0.999, 0.995, 0.990, 0.983, 0.975, 0.964. 0.954, 0.951, 0.950, 0.949

k_η = 0.990, 0.980, 0.970; 0.960; 0.950; 0.940; 0.930; 0.920, 0.910, 0.900

k_γ = 1.050, 1.100, 1.150, 1.200, 1.250, 1.300; 1.350, 1.400, 1.450, 1.500

k_H = 1.000, 0.940, 0.910, 0.880, 0.850, 0.850, 0.790, 0.760, 0.730, 0.700

k_Q = 0.945, 0.940, 0.920, 0.895, 0.870, 0.845, 0.831, 0.825, 0.820, 0.810

The vacuumetric suction height of a hydrotransport installation can be converted from water to slurry as follows:

$$H_B^{sl} = H_B^\circ \,(\,1 + as)\, n + H_a \frac{\rho_{sl} - \rho_0}{\rho_U} - , \text{м} \ \ H2O$$ (6.11)

126

where H_a is the atmospheric pressure, m water, the value of which is presented below.

Height above sea level, m	−600	−200	0	+100	+500	+1000	+2000	+5
Atmospheric pressure, m water	11.3	10.6	10.3	10.2	9.7	9.2	8.1	

Normal operation of a hydrotransport installation requires that the vacuumetric suction height developed by the coal pump H_B^H be greater than H_B^{sl}; therefore, the permissible cavitation reserve Δh is

$$\Delta h = H_B^{sl} \frac{p_b - p_p}{\gamma} + \frac{v^2}{2g}, \text{ м } H_2O \qquad (6.12)$$

where p_b is the absolute pressure at the input of the coal pump, N/m^2: p_p is the vapor pressure of the slurry transported, N/m^2.

t^0 C	0	10	15	20
p_p $10^5 N/м^3$	0,006	0,012	0,017	0,024
t^0 C	25	30	35	40	45
p_p $10^5 N/м^3$	0,032	0,043	0,057	0,075	0,098

The geodetic suction height developed by the coal pump can be determined for water from the expression

$$H_g^0 = H_a - \left(\frac{n\sqrt{Q}}{c}\right)^{4/3} \cdot 10, \qquad (6.13)$$

where n is the rotating speed of the coal pump shaft, rpm; Q is the discharge of the coal pump, m^3/s; c is the cavitation coefficient, which is related to the speed coefficient n_s as follows:

127

$$n_s \quad \ldots\ldots\ldots\ldots\ldots\ldots \quad 50\text{-}70 \quad 70\text{-}80 \quad 80\text{-}150 \quad 150\text{-}200$$
$$c \quad \ldots\ldots\ldots\ldots\ldots\ldots \quad 600\text{-}750 \quad 800 \quad 800\text{-}1000 \quad 1000\text{-}1200$$

For slurry

$$H_g^{sl} = H_a - \varphi \left(\frac{n\sqrt{Q}}{c} \right)^{4/3} \cdot 10, \qquad (6.14)$$

where ϕ is the cavitation reserve factor, which depends on the consistency of the slurry and the speed coefficient of the coal pumps.

The cavitation reserve coefficient ϕ can be taken from table 6.3.

The speed coefficient can be determined by the equation

$$n_s = 3.65 \, n \, \frac{Q^{1/2}}{H_0^{3/4}}, \qquad (6.15)$$

where H_0 is the head of the coal pump, m H_2O.

In order to determine the range of operating conditions of a hydrotransport installation with varying slurry consistency, two sets of curves must be constructed: one $H_H = \phi(Q,s)$, expressing the pressure-flow rate characteristics of the coal pump as a function of discharge and the consistency of the slurry, the other $H_{tp} = f(Q,s)$ expressing the variation in pressure characteristic of the pipe as a function of the same parameters.

Table 6.3

s	Values of n_s					
	70	80	90	100	110	120
0.1	2.50	2.19	1.91	1.72	1.59	1.44
0.2	2.54	2.23	1.95	1.76	1.62	1.47
0.3	2.59	2.27	1.99	1.80	1.65	1.50
0.4	2.64	2.32	2.03	1.84	1.68	1.53
0.5	2.80	2.45	2.15	1.95	1.78	1.62

128

The characteristics of the coal pump and pipeline are expressed in m H_2O; therefore, as the consistency of the slurry increases the pressure characteristics of the coal pump and pipeline are located higher one above the other.

The maximum possible output of the coal pump for operation with slurry

$$Q_{s1} = Q_0 (1 - 0.6 s), \text{ M}^3/hr \qquad (6.16)$$

where Q_0 is the maximum output of the coal pump when operating with water.

For series or parallel operation of coal pumps, the minimum and maximum outputs are determined for the overall characteristics of the coal pumps and the characteristics of the pipeline.

With parallel operation of two centrifugal coal pumps into one pipeline, the operating conditions are determined graphically, as is done for pumps transferring water, considering possible wear of one or both coal pumps during the process of operation. The total output of both coal pumps Q_s should meet the requirement

$$Q_s = (1.7 - 2.0) Q_{opt}, \qquad (6.17)$$

where Q_{opt} is the optimal output at maximum efficiency.

Considering that the variety of coal pumps produced cannot precisely satisfy the requirements for every hydrotransport installation, the catalog is used to select the pump which comes closest to the required characteristics after conversion of the characteristics from water to the slurry which will be used.

If the variation in the operating capacity from the required capacity is too great, the planner has two choices:

change the characteristics of the pump selected by changing the rotating speed of the shaft (by changing the speed of the motor) or decreasing the diameter of the impeller (within permissible limits);

change the diameter of the pipe or the consistency of the slurry.

129

6.1.4. Trends in the Improvement of Coal Pumps

The following main trends have been noted toward increasing the durability and reliability of existing coal pumps or the hydrotransport of coal: improvement of hydraulic parameters of design; creation of wear-resistant shapes of flow-carrying parts; assurance of normal operating conditions.

The main difficulties encountered in the development of high head pumps are those of assuring the necessary suction height; balancing axial forces; assuring reliable operation of seals and development of wear-resistant shapes for flow-carrying parts.

To assure sufficient anticavitation reserve in the input of high pressure pumps, low speed operation of the impeller must be used, leading to a low speed coefficient and corresponding low efficiency, increased dimensions and cost of the pump.

Studies have shown that the head developed by coal pumps, expressed in meters of the column of the fluid being transferred, decreases with an increase in the consistency of the slurry, which is explained by the increase in hydraulic resistance. As the consistency of the fluid transferred increases, the suction capacity of the coal pumps drops; steps should therefore be taken to assure normal, cavitation-free operation.

As the head of a coal pump increases, the unbalanced axial force increases. It can be decreased by placing slot seals of the impeller at one diameter and allowing leakage through the suction side of the impeller. In this case the shaft seal on the delivery side is exposed to the suction pressure. Various unloading apertures in the hub of the impeller and the installation of radial ribs on the impeller disc are also used. All of these methods result in energy losses in a coal pump with suction on one side.

At the present time, VNIIGidrougol' and UrkNIIGidrougol are developing a coal pump with suction on both sides for transportation of small particle materials under a pressure of 2.5 MPa.

An important part of coal pumps for suspended materials, which largely determines the reliability and durability of the entire coal pump, is the seal unit. Increasing the service life of seals requires that as few solid particles enter the seals as possible. In soil pumps, the seal gaps are protected from slurry primarily by creating a counterflow of pure water through the gaps. The gaps between the impeller and armored disc, as well as the impeller gland seal, are protected in this manner. The flow rate of water through the gland seal represents about 1.5% of the quantity of slurry transferred. In order to seal the gap between the impeller and the armored disc on the suction side, a quantity of water equal to approximately 10% of the slurry transferred is required, or if the pump is the second stage of a system, this quantity of water reaches 20%. The head of the pure water must be 10-15 m H_2O higher than the head developed by the pump.

This method of protection of the seals of the impeller from wear requires significant consumption of electric power to force the extra water through the seals, and causes dilution of the slurry.

Another method of decreasing the wear of seals on the impeller is to decrease the pressure beside them. In soil pumps this is achieved on the output side by placing radial blades (impellers) on the outer surface of the impeller disc, forcing particles of solid material out to the periphery by centrifugal force as the impeller rotates and decreasing the pressure near the axis of the impeller.

Studies performed at UkrNIIGidrougol have established that an analogous sealing effect is achieved by increasing the diameter of the impeller disc by 10-15% in comparison to the outside diameter of the impeller vanes. In this case the maximum possible leakage caused by wear of the forward seal is decreased by a factor of 1.7. Therefore, even significant wear of the forward seal has little influence on the operating characteristics of the pump, i.e., the head and efficiency, which can be greatly reduced in pumps with impellers of the ordinary design.

One means for increasing the service life of the moving parts of transport pumps is to design the geometry of the flow-carrying portion so as to reduce the velocity of the flow next to surfaces subject to

wear and most importantly to avoid eddy zones, which are the primary
cause of local wear.

The problem of increasing the wear resistance of pumps transferring
abrasive slurries is being studied by many scientific research organizations.
The methods which they have developed to control wear can be reduced
primarily to the use of special wear-resistant alloys, various types of
antiabrasive coatings, introduction of large numbers of interchangeable
parts to the design of hydraulic machines, which, along with some increase
in the service life of pumps, also achieves a significant increase in
their cost.

Successful solution of the problem of increasing the wear resistance
of coal pumps will require careful study of the processes occurring in
the flow-carrying portion in order to locate and eliminate areas with
high velocities and intense vortex formation. Elimination of eddy
zones and reducing the speed of the flow near the walls which limit the
flow can increase service life still more. The method of designing
the flow-carrying portion by studying the structure of the flow on an
aerodynamic test stand, which allows the essential nature of the phenomena in question to be studied most deeply, is particularly effective
in this process.

UkrNIIGidrougol' has created an aerodynamic test stand and used
it to design the flow-carrying portion of the 12U10 and 10U5 coal pumps.

In the 12U10 coal pump, the structure of the flow in the suction
pipe was studied in order to find wear-resistant shapes for the intake
section of the coal pump. It was established as a result of these
studies that the reason for intensive wear of the suction pipe of the
12U10 is the great swirling of the flow in this section. The circumfer-
ential velocity of the flow is as high as 10-20 m/s, depending on
operating conditions. This results from leakage of fluid with some
initial swirling through the front seal of the impeller, and also from
the backflow from the impeller into the suction pipe under light-load
conditions. In order to eliminate the causes of rapid wear of the

suction pipe, it was necessary to decrease the influence of backflow
and leaks on the flow. This was achieved by installing a guide which
directs the leaks to the impeller parallel to the intake edge of a
vane and reflects backflow from the impeller under light-load conditions.
The use of this leak guide also increased the efficiency of the coal
pump by 4% and improved its cavitation qualities.

A study of the flow structure in the throat of the 12U10 coal
pump with and without the leak guide showed that the flow in the throat
is not symmetrical relative to the impeller. The difference between the
minimum and maximum values of static pressure at the same time is
between 10 and 20%. The imbalance of pressure and speed in the throat
results from the large gap between the impeller and the outlet scroll so
that even under optimal operating conditions the pressure around the
impeller is not uniform.

The studies performed were used to design the 12U10M coal pump,
which, according to the results of industrial testing of an experimental
model, has an efficiency 3-10% (depending on operating conditions) higher
than that of the 12U10 coal pump, even with significant wear of the
front seal (figure 6.3).

The design of the impeller of the 12U10M features enlarged discs.
An experimental production run of these pumps has passed industrial
testing in the hydraulic mines of the Donets basin. The results indicate
that the 12U10M coal pump has longer service life, more stable operating
characteristics and higher efficiency than the series produced 10U10
coal pump. It should be expected that the modernization of other soil
pumps to produce pumps like the 12U10M should also have a positive effect.

One important trend in the creation of high pressure pumps is the
production and use of wear-resistant materials. Wear resistance is
determined primarily by proper selection of the material of which a
pump is made. Studies have shown that rubber has significantly (at 8 to
10 times) more wear resistance than standard type 3 steel. Rubber is
quite promising as a material for lining the casing and impeller of
pumps transferring particulate materials in finely ground form. For

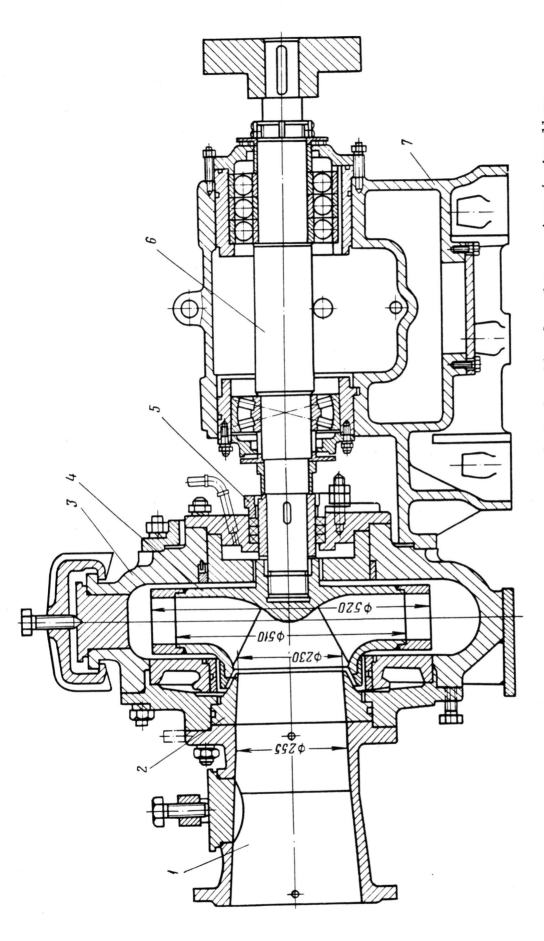

Figure 6.3. Design of 12U10M coal pump: 1--suction pipe; 2--leak guide; 3--volute casing; 4--impeller; 5--seal; 6--shaft; 7--frame.

134

coarser materials, hard alloys, cast stone and other materials must be used.

High-chromium type "SS" cast iron is widely used. With proper casting and heat treatment, this material can be highly wear resistant. One shortcoming of this metal is the difficulty of mechanical working, another is its brittleness, resulting in tthe failure of impellers and discharge pipes if lumps of metal or large, hard inclusions are present in the slurry.

VNIIGidromash Institute recommends that cast iron type IChKh28N2 be replaced in soil pumps by cast iron type IChKh16MT or steel type 40 KhGSNL and 35KhNVFL, which are significantly easier to work and have greater wear resistance, combined with satisfactory flexural strength. However, these steels require precise heat treatment, creating certain difficulties at manufacturing plants. Therefore, assurance of accurate heat treatment conditions is an important task for increasing the wear resistance of coal pump parts.

Another important area is that of improving the operating conditions of pumps, achieved by improving the technology of hydrotransport and by the creation of supplementary devices to protect the machines from mechanical damage and stabilize their operating conditions.

6.2. Piston Pumps

The advantages of piston pumps over centrifugal pumps for mainline hydrotransport are primarily that they create high head during startup of the installation or in case the resistance to slurry flow increases due to a change in consistency and assure normal operation of the entire installation (table 6.5). They also have comparatively high efficiency, are reliable and durable in operation.

Type U8 piston pumps are manufactured for pumping of drilling fluids, frequently containing large quantities of sand. The service life of the pumps as a whole, including their most important body parts, should be no less than 15 to 20 years, according to the manufacturer. The pumps are designed to allow the replacement of broken or damaged elements.

135

The bearings should last at least 20,000 hours, and the efficiency of the pump is no less than 80 to 85%, which is achieved by careful and accurate manufacture and assembly of all parts of the pump and efficient design of the seals. The interchangeable parts of the pump are easily accessible for replacement when they are worn.

Piston pumps were tested by GrozNII and UkrNIIGidrougol' to determine their suitability for transportation of coal slurries of various consistencies with various particle sizes. It was found that they can create high pressure and can transport high consistency slurries of finely ground coal. The operation of piston pumps is particularly reliable for hydrotransport over long distances of water-coal suspensions consisting of highly concentrated and dispersed systems with particle diameters not over 0.2 mm.

Table 6.4

Pump	Operating pressure, MPa	Discharge, m^3/hr	Hydraulic power, kW
U8-3	5.5-15.0	162.0-61.0	370
U8-4	9.5-20.0	128.0-58.5	440
U8-7	18.5-32.0	124.0-72.0	736

When transporting slurries consisting of 0-3 mm particles of coal, even at high consistencies, up to 50%, the pump valves operate normally. At times, sudden seating of the valves was observed, though it was not accompanied by knocking. The operation of the pump with high consistency slurries causes a decrease in discharge and in volumetric efficiency from 92% when operating with water to 82%.

Normal operation of the valves can be assured only if the slurry contains no lumps of coal larger than 3 mm in diameter.

At high slurry consistencies, coal particles find their way between the turns of the valve springs, resulting in a decrease in the load on the valves and a reduction in their lift height.

Piston pumps transport high consistency slurries reliably and stably; however, they do not meet the requirements placed upon them completely: they yield comparatively low flow rates, requiring parallel connection of several machines, making pumping plants more complex and expensive. When used individually, piston pumps are suitable only for projects with relatively low throughput, on the order of 500,000 tons of solids per year. The pressure created by these pumps (over 15-20 MPa) requires the use of expensive pipe and is significantly higher than the optimal pressure. Their discharge is usually not over 180 m^3/hr; they require complex transmission systems for connection to electric motors, as well as large numbers of qualified servicing personnel. They create a pulsating flow of slurry and limit the particle size of the material which can be transported.

According to catalog information, the ratio of pump mass to power produced for series produced piston pumps is 0.08-0.10 kg/kW; therefore, the mass of a piston pump for mainline hydrotransport with a flow rate of 1000 m^3/hr and a pressure of 6-8 MPa is over 60-80 tons. The large dimensions and mass of piston pumps limit their area of application.

The need has arisen to plan new piston pumps meeting the requirements of mainline hydrotransport. One attempt at the creation of a piston pump with a discharge of 200-500 m^3/hr at a pressure of 5-6 MPa was undertaken by UkrNIIGidrougol'. Other organizations have also undertaken the task of creating piston pumps with greater throughput than series produced pumps.

6.3. The NPP Pump

Recently, special type NPP feeder pumps have been developed, in which some of the shortcomings of piston pumps have been eliminated: no cumbersome power transmissions and no contact between slurry and moving parts. The experience of operation of piston pumps indicates that they are suitable and durable for the transfer of coal slurries (actual

137

service life of pumps as great as 15,000 hours).

Type NPP pumps developed by UkrNIIGidrougol', in which the slurry is separated from the operating parts of the pump by means of a water plug are particularly suitable for the hydrotransport of highly abrasive materials.

A piston feeder pump is distinguished by the fact that the abrasive solid particles do not pass through the pump, and therefore, this type of pump can be used for the hydrotransport of highly abrasive materials, for example, ashes and slags for hydraulic stowing.

A diagram of a type NPP-2 piston feeder pump is shown in figure 6.4. The pump consists of two cylinder units, floating pistons, a distributor valve system and valve shifting mechanism, loading pipes, as well as intake and exhaust valves.

High pressure water is fed to the NPP-2 pump by a centrifugal pump. The water shifts the pistons alternately from one extreme position to the other. The cylinders are located coaxially in the blocks, their pistons are connected by shafts; each cylinder consists of driving and operating cavities. The water is distributed to operate the pistons by two three-position distributors connected to the valve switching mechanism. The pistons stop briefly in the extreme positions, allowing the valves to seat without hydraulic shock.

The hydraulic system is designed so that the volume of the loader pipes is somewhat greater than the swept volume of the cylinder; therefore, a hydraulic plug remains in place, protecting the cylinders from contact with the slurry.

The feeder pump is started by filling it with water to remove air and to form the liquid pistons. After the air is removed from the hydraulic system, the process of suction of the slurry into one loading pipe and discharge of the slurry from the other loading pipe begins.

The large valve openings allow slurries containing lumps of solid materials up to one-third the diameter of the pipe to be transferred.

The NPP-2 pump was tested for hydrotransport of slurries of anthracite, ore and other materials.

The pump is driven by a centrifugal pump; therefore, the characteristics of the NPP-2 are not rigid like those of piston pumps, but rather soft like those of centrifugal pumps.

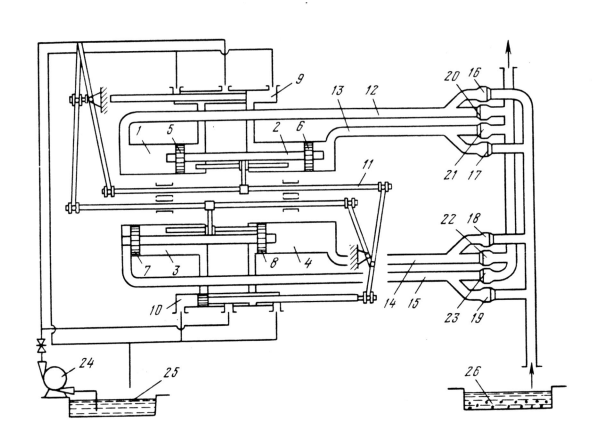

Figure 6.4. Diagram of pump-feeder: 1-4--cylinders of cylinder block; 5-8--floating pistons; 9-10--valve distributors; 11--switching device; 12-15--loading pipes; 16-19--intake valves; 20-23--exhaust valves; 24--driving pump; 25--water reservoir; 26--slurry reservoir.

It has been found that the parameters of the pump are practically independent of the density of the material transferred. An experimental model of this pump was tested in 1972 at the Balakleysk cement and slate plant, transferring a cement raw materials slurry with a density of up to 1500 kg/m^3 over a distance of 9.2 km. The tests indicated that the design of the valves needed improvement.

Technical Characteristics of NPP-2 Pump

Discharge, m^3/hr	800
Discharge pressure, MPa	9
Number of cylinders	4
Cylinder diameter, mm	600
Double strokes per minute	10
Dimensions, mm	
Length	13,525
Width	3,045
Height	3,145
Mass, kg	33,700

6.4. Airlifts

Single-stage and multiple-stage airlift hydraulic hoist systems developed by Donetsk Polytechnical Institute have been in operation for a number of years in hydraulic mines in the Donets basin.

The operating principle of an airlift is as follows: air is fed into the bottom end of a pipe immersed in slurry and open in two directions. The bubbles of air rise upward through the pipe, drawing along the water and the solid particles which it contains. Slurry will move in an airlift only if the quantity of air fed into the pipe is sufficient to create the necessary lifting force.

The slurry is fed to a mixer, which also receives compressed air. The compressed air and slurry interact in the mixer, and the three-phase medium, consisting of air, water and the solid material, rises to the surface.

Slurry is fed into the first stage of an airlift as follows. The slurry flows down flumes from the mining sections to the sump portion of a shaft 100 m deep. The compressor operates at 0.8 MPa. The slurry is drawn up from the lower portion of the sump by the airlift and fed directly to the surface by a single-stage airlift or to the next stage of a multiple-stage airlift, from which it is fed from stage to stage until it reaches the surface.

Reliable operation requires that the mine have a standby airlift of equal throughput capacity.

The airlift installation includes emergency and regulating containers made so that the solid materials which they hold does not sink to the bottom of the containers, but rather drops into the sump. In case of an unplanned shutdown of the airlift, the slurry returns to the sump. The mass of solid material which precipitates is carried to the surface when the airlift is restarted by the airlift itself, without additional mechanisms.

All mine inflow in this case is removed by the airlift, which thus serves both as a hydraulic hoist and main mine drainage pump.

An airlift hydraulic hoist working with a sump allows reliable regulation of the throughput of the hydraulic hoist due to the great difference in possible levels of the slurry.

In a single-stage airlift, the maximum throughput of the airlift installation corresponds to conditions such that the level in the sump is at the maximum. The throughput of multi-stage airlifts can also be regulated over a wide range.

DPI has developed this and other methods of feeding an airlift with vertical containers and with dumping of the slurry from ordinary slurry collectors into an airlift intake pipe. Each of these systems can be used under the proper conditions.

Air can be supplied for the stages of a multi-stage airlift either individually to each stage or from a common air collector. One of the most important conditions of normal operation of an airlift is proper selection of the mixer immersion depth.

The immersion depth is usually determined by the equation

$$h = \frac{10p}{\delta_S}, \text{ м},$$ (6.18)

where p is the maximum air pressure supplied by the compressor, in MPa; δ_s is the relative specific gravity of the slurry.

The necessary immersion depth can be achieved by deepening the shaft, or if this is impossible a coal pump plus airlift hoist system is used. The airlift should operate with a relative immersion of approximately 0.5, the most economical depth.

One important element of an airlift is the mixing chamber, in which the compressed air and slurry are mixed together. Various designs of mixers are used: with central or lateral location of the air pipe. Airlifts with lateral location of the air pipe are becoming increasingly common in the mining industry.

Complex processes of mixing of slurry and air, not suitable for strict mathematical analysis, occur in an airlift. It has been found experimentally that the design of the mixing chamber and speeds of the flows of gas and liquid have a great influence on the flow process in an airlift. The throughput of the airlift, DPI recommends, should be calculated by the equation

$$Q = k_c \ D^{2.5}, \ \text{M}^3/\text{hr} \qquad (6.19)$$

where D is the diameter of the lift pipe, cm; k_c is the flow rate coefficient.

Operating experience indicates that an airlift installation must operate strictly according to its standard technological conditions. Under normal conditions, the sump portion is filled with water, and sumps and water collectors may be drained to their lowest level or filled with water as the inflow from the mine arrives.

When a multi-stage airlift is started, the air is first fed into the upper stages, then into the lower ones. During the process of operation, the level of fluid in the sump falls, and the throughput in the stages of the airlift is automatically decreased. If the inflow of slurry increases, the level of slurry in the sump rises and the throughput of the airlift installation increases, i.e., the system is automatically self-regulating.

The airlift installation is shut down as follows: first the coal pumps which transfer the slurry are switched off, then when the "cutoff level" is reached in the sump, airflow to the bottom stage is interrupted, followed by the higher stages.

Airlifts usually operate jointly with coal pumps. The slurry transferred by the airlifts must be transported to a beneficiation plant or other consumer, which can be done with coal pumps or by gravity flow.

Airlift hoist in the Donets basin is used to raise four to five thousand tons of coal per day to the surface from a hydraulic mine; the planned capital investment for construction of the hoist was 1.5 million rubles, operating cost per year 0.73–0.76 million rubles, including 0.63–0.67 million rubles for electric power.

The selection of effective hydraulic hoist equipment should be performed considering the cost of construction and operation of the installations, as well as the cost of crushing of the coal. Since various hydraulic hoist methods have been developed to varying degrees, the economic calculations are quite approximate in nature.

According to Dongiproshakht Institute, which has compared three types of hydraulic hoists (coal pump, airlift and hopper-feeder hoists) the coal pump hydraulic hoist method is most favorable from the standpoint of capital investment. VNIIGidrougol' also concluded that coal pump hydraulic hoists are less expensive to operate than airlifts.

6.5. Feeding Apparatus

Hydrotransport of particulate materials in a stream of liquid requires that the flow have a certain reserve of energy. Various methods can be used to impart energy to a flow. The most common scheme is to utilize a centrifugal, volumetric, screw or other type of pump. In the pump, the slurry is given additional energy, which is required to move it through the pipe. This is the simplest scheme, but it still has a number of shortcomings -- severe abrasive wear of the pump through which the slurry passes and great particle size reduction of the materials.

These shortcomings can be eliminated to some extent in devices in which the particulate material is placed in the pipe so that it bypasses the pump. The pump, operating in clear water, feeds the water into the pipe. Immediately downstream from the pump a device is included allowing the particulate material to be introduced to the pipe, after which it is transported by the energy imparted to the water by the pump. A diagram of such a feeding apparatus is shown in figure 6.5.

A special valve device directs the stream of water supplied by the pump alternately into a number of branches (two shown in figure 6.5). The particulate material is charged into branch I, with valves 5,6, and 7 closed. Valves 2,3, and 4 are in the position shown in the drawing. This causes the solid material from branch II to be washed downstream.

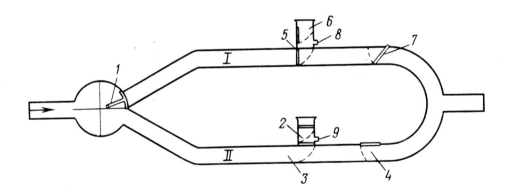

Figure 6.5. Diagram of feeding apparatus: 1--switching valve device; 2-7--valves; 8-9--feeding pipes.

Feeding apparatus is usually divided into two classes according to operating principle. Class one includes feeding apparatus in which a volume of water equal to the volume of the material loaded is removed from the pipe as the material is loaded into the pipe. Exchange-type apparatus requires the removal of waste water, frequently containing significant quantities of finely dispersed solid particles. In estimating the energy characteristics of feeding apparatus of this class one must consider the energy loss resulting from the need to pump away the waste water. The solid particles receive the energy of a stream of pressurized water in these devices.

The feeding apparatus used in systems which transport slurries of high consistency has relatively low efficiency; also, when these devices must be used in series, as is required for long distance hydrotransport, the slurry must be thickened at each intermediate pumping plant, removing all excess water, since the concentration of the solid particles will be very low by the end of the transportation cycle.

Feeding apparatus is also unsuitable for the transportation of fine-particle coal or slurries, since the waste water produced contains large quantities of finely dispersed particles. Clearly, feeding apparatus can be used for long range hydrotransport only in exceptional cases.

The second class of devices includes apparatus which force the solid material into a stream of high-pressure water. These devices do not displace water, but rather the energy is imparted to the material to be transported as it is loaded into the pipe. These devices include screw, piston, jet pump and other types of devices.

In order to force the solid material into the high pressure pipe, the pressure created by the feeding device must be higher than the pressure of the water in the pipe.

In screw type devices, the material itself must form a plug which prevents the liquid from escaping from the pipe. Under certain conditions, this can be achieved at comparatively low pressures (0.3-0.5 MPa), given the proper particle size of the particulate material. If the pressure is over 0.5 MPa or the particle size of the coal decreases, the screw device cannot retain the plug and it ceases to operate as a pump.

Pipe feeder apparatus, which can be easily placed in mine openings, is particularly promising for hydraulic mines. VNIIGidrougol', UkrNIIGidrougol' and other institutes are working on the development of pipe feeders. Pipe feeders are more widely used abroad (in Japan) than chamber feeders due to their smaller dimensions, simpler design and relatively low cost.

Chamber type feeding apparatus used for the feeding of run-of-mine slurry has the shortcoming that the potential capacity of the chambers is only 10-20% used, and after the flow of slurry joins the flow of pressurized water, the resultant concentration of slurry is so low that hydrotransport is uneconomical. These devices can be used where the chambers can be loaded with dry large lump coal, particularly highly abrasive types. They can be used for the transportation of stowage materials in mines, for hydrotransport of crust rock in open-pit mines, for transportation of coal from shaft mines using dry mining technology to beneficiation plants, etc.

Sibgiprogormash has developed the AEV-2 hydrotransport system, consisting of a feeding apparatus, rock preparation section, pumping plant, water collector and pipe system. The apparatus utilizes sluicing and forced washdown. The maximum operating pressure is 7 MPa, rock throughput 180 t/hr, chamber capacity 0.6 m^3, number of chambers 2.

The apparatus features special gates which reliably seal the chamber. The spherical gates have large open cross sections, and their design prevents the formation of dead "spaces" within the chamber.

A drum-type feeder designed by the Yasinovatskiy machine building plant can be used to feed solid materials into pressurized pipes for hydrotransport of coal and rock over distances of up to 4 km. The body of the feeder has two fittings for the input of slurry into the pipe and for drainage of water from the individual cells in the drum after the solid material is washed out. The pipe to be used for input of slurry and drainage of water is connected to the feeder using quick-split joints. There is a hatch on the discharge fitting to allow removal of the solid material in case of plugging of the pipe. The feeder is charged by means of a rotating drum with seven cells each 150 mm in diameter; the overall drum diameter is 450 mm. Water is drained from the cells through four openings drilled in the spaces between cells. The drum rotates between two covers. The cells pass beneath a loading funnel, where they are filled with solid material. The cells filled with solid material are then connected to the pipe fittings, the material is washed out into the pipe, and the cell is returned for reloading.

Since the rotating drum stops the water flow as each new cell is rotated into position, the pressure in the water intake and discharge pipes pulsates as a function of the speed of flow of the water in the slurry line. Air caps can be used to damp these pulsations of pressure. The throughput of the feeder is 40 tons per hour of coal, 75 tons per hour of rock. The operating pressure of the pump is up to 1.5 MPa, maximum particle size of material 100 mm, electric motor power 8 kW.

The drum feeder is small (height 1250 mm, length 1680 mm, width 840 mm) and light (2060 kg). Its operation is reliable. The end covers of the body can be surfaced with special alloys to increase durability, and their thickness can be selected depending on the desired service life.

Long term operation of a drum-type feeder in the laboratory of UkrNIIGidrougol' indicates that it can be used successfully as a sampler or for hydrotransport of coal and rock.

Feeding apparatus may be reliable and highly effective when used in medium distance transport systems (up to 8-10 km) for highly abrasive solid materials, particularly when the solid material arrives in relatively small quantities and must be stored in an intermediate container. The feeding apparatus combines the function of the intermediate container and transporting device.

There are as yet no good automatic feeding devices. The best system of feeding devices is probably that developed by B.M. Shkundin [12]. The throughput of the apparatus as solid material is 300 m^3/hr, as slurry 2500 m^3/hr, maximum particle size of transported material 200 mm, chamber capacity 15 m^3, cycle duration 550 s. The loading apparatus operates with a head of not over 100 m, the water from the chamber is drained directly into a river (this apparatus was used in the construction of the Hurek regional electric power plant for relatively high concentrations - up to 0.25 by volume).

Under shaft mine conditions the feeding apparatus must operate
jointly with a pump to carry away the dirty water. Pumping of highly
contaminated water requires the use of special pumps, making the job
more difficult.

The coal industry presently utilizes large capacity feeding devices.
A hopper-feeder utilizing the DonUGI design has been in operation for
over 10 years at "Kapital'naya" shaft mine number 6 ("Donetskugol'"
production union). The throughput of the installation is up to 400 tons
per day of rock, which is brought up from a mine from a depth of 320 m
and transported to the surface to a hydraulic tailings area up to 2 km
distant. Long term operation of the hopper-feeder indicates that it is
reliable and durable.

The GIG chamber feeder has been in use in Poland in shaft mines
for hoisting of coal since 1957. This is a two-chamber apparatus
consisting of upper charging gates, worm-type measuring devices, hydraulic
relief valves, a screw feeder and an electric-hydraulic control system.
The chamber volume can be varied, since the chambers of the apparatus
are made up of cast steel sections with an inside diameter of 1200 mm.
The plate-type charging gates are hydraulically driven and include
devices for water washing. As the gates are closed, the sealing surfaces
of the rubber sealing collars are washed.

The area of application of feeding apparatus is limited as to
throughput, pressure, mean particle size and particle size distribution
of the solid material.

6.6. Hydraulic Elevator (Jet Pump)

The operating principle of a jet pump is based on the fact that the
energy of the operating fluid in the mixing chamber is imparted to the
slurry which is drawn in, the streams of these two fluids are mixed,
then transported for some distance.

In a jet pump (figure 6.6) water is fed under high pressure into a nozzle, then from the nozzle is exhausted into a mixing chamber, creating reduced pressure due to its high velocity. This causes the slurry to be drawn in through the suction pipe into the body of the jet pump. In the mixing chamber, the streams of water and slurry are mixed, then fed into the diffuser and the discharge pipe.

The simple design of the jet pump, the lack of moving parts, its high reliability and suitability for transportation of highly abrasive materials, as well as its capability for washing itself out make this device irreplaceable under certain operating conditions.

Jet pumps are used as independent devices for the transfer of various slurries and for supplementary purposes such as the removal of air during startup of centrifugal pumps. Jet pumps are widely used in various areas of industry; in the coal industry, they can be used jointly with centrifugal coal pumps and for various supplementary purposes (for example, to clean out sumps).

The less the angle between the axis of the nozzle and the axis of the suction fitting, the more effective the jet pump. The distance between the nozzle and the intake to the mixing chamber (1) greatly influences the effectiveness of operation of a jet pump. This distance can be taken in approximate calculations as

$$l = 2d_0,$$ (6.20)

where d_0 is the diameter of the nozzle.

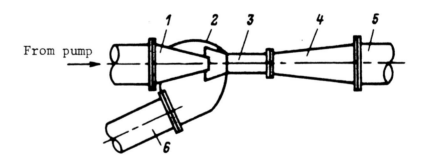

Figure 6.6. Diagram of a jet pump: 1--nozzle; 2--body; 3--mixing chamber; 4--diffuser; 5--discharge pipe; 6--suction pipe.

The length of the mixing chamber should be such that the process
of mixing of the two flows is completed within the chamber

$$l_3 = 4{,}65 \, d_0^{0,2} \; d_3^{0,8} \, ,$$

(6.21)

where d_3 is the diameter of the mixing chamber (a cylindrical mixing
chamber is recommended).

Mixing chamber nozzles are made interchangeable with a Laval,
conical or cylindrical-conical profile. The recommended diffuser
profile is complex, consisting of three parts.

A jet pump is described by the head coefficient, suction coefficient,
etc.

The head coefficient

$$a = \frac{2g H_{jp}}{v_2^2} \, ,$$

(6.22)

where v_2 is the speed of the liquid at the intake to the mixing chamber;
h_{jp} is the head developed by the jet pump, equal to the sum of the
geometric suction height, geometric discharge height and the loss in
the suction and discharge pipes.

The suction coefficient

$$\beta = \frac{\gamma_s Q_s}{\gamma_0 Q_0} \, ,$$

(6.23)

(the subscript "0" relates to the water, the subscript "s" to the slurry).

Figure 6.7 shows the function $a = f(\beta)$ for common types of jet pumps.

Depending on the purpose of a jet pump, various approaches can be
taken to evaluation of its efficiency. If the flow rate of slurry
drawn in by the jet pump is most important, its efficiency is

$$\eta_s = \frac{\gamma_s Q_s H_s}{\gamma_0 Q_0 H_0} \, .$$

(6.24)

150

The efficiency of a jet pump can be expressed through the head and discharge coefficients as follows:

$$\eta_s = \alpha \beta. \qquad (6.25)$$

Jet pumps currently in use have efficiencies of 10-20%; theoretically the efficiency could reach 30%. Low efficiency remains a major short-coming of jet pumps, limiting their area of application.

6.7. Instruments, Fittings, Pipes

The hydraulic mining and hydrotransport of coal involves installations which require monitoring of their operation and control of the process based on certain important parameters: the pressure and flow rate of the water and slurry and the density of the slurry. The pressure and flow rate of the water are measured and regulated using the instruments used in the water supply system of an industrial enterprise. Some modern designs of instruments for slurry are described below.

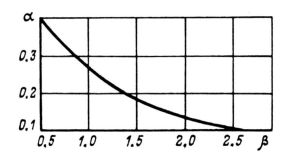

Figure 6.7. The function $\alpha = f(\beta)$ for jet pumps.

Pressure measuring devices. In addition to ordinary process manometers, pressures are measured using transducers of various designs to transmit pressure signals to a control panel or secondary indicators. Common types of devices include diaphragm instruments which have nonlinear characteristics, and electromagnetic sensors, which are used for the measurement of pressures with accuracies of 3-5%. Tensometric type DS-2 sensors with linear characteristics, of accuracy class 0.5, designed for measurement of pressure in all media (water, slurry, concrete cement mass, etc.) have been used in the USSR for measurement related to engineering problems (figure 6.8) [18].

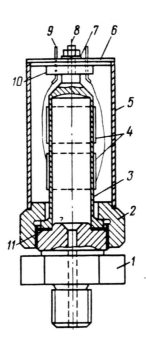

Figure 6.8. Pressure transducer: 1--nipple; 2--nut; 3--cup; 4--resistance strain gauge; 5--protective cover; 6--cap; 7--washer; 8--cup attachment nut; 9--strain gauge lead; 10--insulating plate; 11--sealing plate.

Flow rate meters. The most widely used flow rate meter design in practice is based on the Venturi tube principle. However, it is difficult to convert the pressure to an electric pulse and to prevent plugging of pressure measurement hoses. Two Venturi tube systems using the tensometric principle of pulse conversion have been suggested, based on direct attachment of strain gages to the body of a flow rate measurement or the use of cup-shaped sensors (figure 6.9). The sensing elements in

152

these systems are protected from the medium in the pipe. The surface is finished to class 5-7 smoothness and resistance strain gauges are glued to it (gauge length 50-70 mm, resistance 180-220 Ohms).

The flow rate meter shown in figure 6.10 is recommended for pipe more than 200 mm in diameter.

In both types of flow rate meters, the pulse is fed to amplifier UI and then on to remote instrument VP. Both of these instruments are non-inertial and have linear characteristics during both loading and unloading. These flow rate meters can be used at any pressure with any conditions of movement. Figure 6.11 shows a strain gauge speed measuring device for finely dispersed media designed by the Fedorov Institute of Mine Mechanics (IGM i TK). The resistance strain gauge is glued to a flat flexible plate and protected by a rubber shell. The head and the signal across the strain gauge change as the speed changes.

Hydraulic mining, hydrotransport and water supply of hydraulic mines place special requirements on pipe fittings. The fittings, depending on the nature of movement of the sealing element in the valves, can be divided into gate valves, in which the sealing element moves perpendicularly to the axis of the sealing surfaces in the valve body; globe valves, in which the sealing surface moves along the axis of the valve seat; and cocks, in which the sealing element rotates around an axis displaced laterally from the sealing surface of the valve body.

The main type of sealing fitting used in hydraulic mines is the gate valve. All fittings must be universal, i.e., used in the main drain lines of shaft mines with ordinary mining technology and in hydraulic mines. The operation of gate valves should not require clear water washout, since this frequently complicates the design. Valves should generally be hydraulically driven, though the use of electric or manual drives is also possible. The sealing surfaces should be made of hard alloys, since they are the most reliable materials and resist fracture.

A metal and rubber seal works quite well when suspended particles are present; however, the compressive force must not exceed a certain maximum, since otherwise the rubber will be damaged. Plug cocks are used with suspended particles, but have not become common in the coal industry as yet.

153

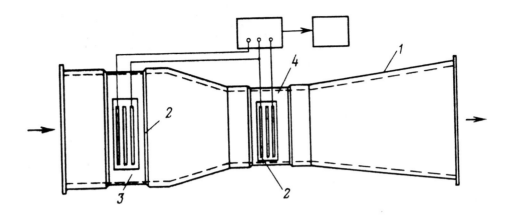

Figure 6.9. Contactless resistance strain gauge flow rate meter:
1--diffuser; 2--resistance strain gauge; 3--confusor; 4--constricted
portion of Venturi tube.

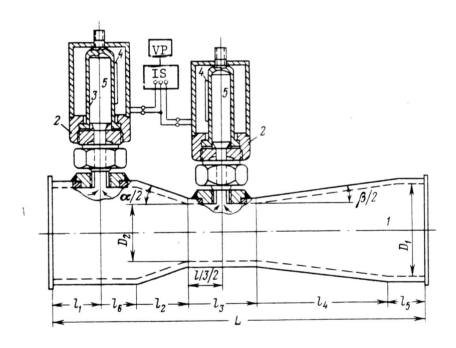

Figure 6.10. Diagram of Venturi flow rate meter with pressure sensors:
1--Venturi tube; 2-3--pressure sensor and sensing organ; 4--primary and
compensation resistance strain gauges; 5--petroleum jelly

Check valves cannot operate reliably with slurry, while valves operating with contaminated water must have elastic sealing elements.

Safety valves operate unreliably with slurry, since lumps of solid material may find their way beneath the valve; therefore, they should be protected from entry of lumps into the sealing unit.

Air escape valves are intended to bleed air from the system and should be designed to prevent entry of solid materials into the sealing devices. It also should be possible to seal the valves.

Standard types of high pressure pipe fittings and pipe elements for hydraulic mining and hydrotransport have been developed. The most important parameter of the series is the size of the standard passage, 100-450 mm for all types of fittings.

The standard calls for the manufacture of steel pipe with inside diameters of 100-350 mm, operating at pressures of up to 10 MPa for the water lines of hydraulic mines. Pipe for hydraulic drives, tees with quick-split joint flanges, taps and adaptors are manufactured with standard inside diameters of 100-450 mm, maximum pressure 10 MPa. Pipes of all types have welded flanges for quick-split joints.

Series ZG gate valves, designed for pressures of up to 10 MPa, with standard passages of 100-300 mm, as well as series ZU gate valves for water lines and slurry lines in hydraulic mines with standard passages of up to 400 mm, are manufactured for use in hydraulic mines.

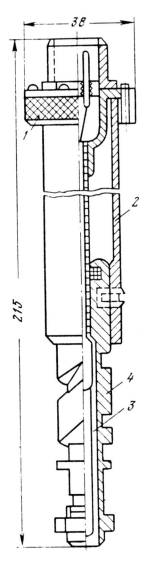

Figure 6.11. Diagram of a strain gauge speed measuring device: 1--moving cover; 2--sliding nut; 3--resistance strain gauge; 4--chamber in body of sensor

155

CHAPTER 7. UNSTABLE OPERATING MODES IN THE HYDROTRANSPORT OF COAL

7.1. Transitional Operating Conditions in Hydrotransport Systems

Transitional conditions in hydrotransport systems occur during startup, shutdown, upon changes in density of the fluid, and in emergency situations (sudden shutdown of the hydrotransport system or intermediate stations in multi-stage slurry transport systems). Variations in pressure and speed increase the wear of pipes and equipment; therefore, the task at hand is to select the placement of pumping plants along the pipeline and determine the best conditions for their operation, startup and shutdown.

The calculation of the pressure pulsations in fluid transferred by piston pumps is very complex and yields inaccurate results. Simple equations have been suggested by the "Parseisen Contororudziu" Company, based on the assumption that the pump valves operate ideally; the axial lines of the cylinder and crank shaft are at right angles; there is no resonance and the hydraulic efficiency is 100%.

For suction units

$$P_p = 3{,}5 p^{0{,}5} v_p z \left(\frac{B}{d_n} \right)^2; \quad z = 1{,}22 \left[\left(\frac{c}{d} \right) \left(p_d - p_s \right) \right]^{\frac{1}{2}}; \tag{7.1}$$

for discharge units

$$v = \pi S n;$$
$$z = \left[\left(1 + \frac{c}{d} \right) \left(p_d - p_s \right) \right]^{\frac{1}{2}}, \tag{7.2}$$

where P_p is the pulsation of pressure from peak to peak; $p = \gamma/g$ is the density of the fluid; v_p and B are the speed and diameter of the plunger; z is a factor; S is the stroke of the piston; n is the number of revolutions of the drive shaft per minute; d_n is the diameter of the suction and discharge apertures in the pump; c is the volume of the suction cavity; p_d, p_s are the pressure during discharge and suction; d is the displacement.

It is erroneously thought that a decrease in pressure pulsations

156

yields a less pulsating flow (pressure pulsations propagate through the fluid at the speed of sound and are independent of flow).

Various dampers are used to damp pressure fluctuations. Flexible quick-split couplings satisfy the broad demand for inexpensive compensators, used for compensation in both the axial and transverse directions at low pressures (on the order of 5 MPa). They have a long service life, decrease impact pressure by 15%, and completely eliminate pulsating pressure fluctuations.

To protect pumps and fittings from pulsation wear at the intake and outlet of repumping plants, one need only install quick-split couplings before and after the pump.

A diagram for damping pressure pulsations is shown in figure 7.1. The chamber contains an elastic envelope filled with carbon dioxide or air, the pressure of which is adjusted to the operating pressure of the pump. The device is installed on the discharge pipe downstream from the pump. Pressure pulsations in the pump are damped by the compressibility of the gas or air in the envelope. This pressure compensation effect can be modeled by a system which reduces the pulsations of voltage by means of a compensator. This device is particularly effective at low frequencies (up to 10 Hz).

A design with a coating which contracts along the length of the pipeline, or a device using the elasticity of a solid (for example, rubber without an air layer) can be used to eliminate high frequency pulsations. These types of dampers are ineffective for low frequency pulsations.

Bladder-type structures can be used to eliminate pressure pulsations at the rotating frequency of the pump (60 to 200 rpm). For high-speed pumps, high-

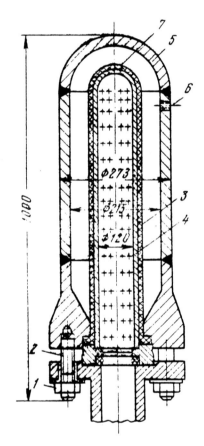

Figure 7.1. Pressure pulsation damping unit: 1--M3 nut; 2--studs 120 mm in length, 8; 3--rubber shell; 4--pipe; 5--metal cap; 6--air feed adjustment screw; 7--flexible rubber mass and air

order pulsations can be eliminated by the use of a damper in which the compressible body is protected from wetting by the fluid transferred by the pump. When a sealed gas is used, its pressure must be tested frequently and the required gas pressure maintained in the envelope; otherwise, the pressure damper will not operate effectively.

A liquid pressure pulsation damper used by the "Palsko" Company takes consideration of the fact that sound propagates through liquids at speeds significantly higher than the speed of propagation of sound in gases. Therefore, at the same frequency the amplitude of pressure pulsations in liquids is significantly greater. The pulsation wavelength must be considered in planning of pressure pulsation dampers. The wavelength becomes greater for liquids; consequently, dampers capable of attenuating low frequency waves must be used. Π filters are used to damp high frequency waves; however, filters which damp low frequency waves are also needed.

The "Palsko" Company pulsation dampers are designed for the use of ordinary plunger pumps, with a shaft rotation speed of 100-300 rpm. For high speed piston pumps (2000-4000 rpm) the wavelength decreases; therefore, dampers are designed using formulas for filters with low throughput capacity, called acoustical filters. These filters are used in axial plunger pumps and are designed by computer. The initial design data are determined from equations (7.1) and (7.2).

The length of the pipe from the pump to the intake of the device should not exceed a/44 or, where a = 800 m/s, on the order of 20 m (here a is the propagation velocity of the pulsation waves). The length of the pipe following the damper is also important; it should be longer than the pipe before the damper.

Fluctuations of the pressure in elbows and pipes, pressure jumps during startup and shutdown influence the pulsation wear of pipes, fittings and equipment. The pressure of pulsation or hydraulic shock produces a reaction force at each bend. The action of this force on the wall of the pipe or elbow can be determined by the laws of momentum applied to the massive fluid in the volume equal to the space in the pipe at the bend.

Since the change in momentum per unit time is equal to the sum of the projections of all forces acting on this mass of fluid on the x axis, then [14]

$$m_1 v_1 - m_2 v_2 \cos \beta = R \cos (90 - \beta) \Delta t = R \sin \beta \, \Delta t, \qquad (7.3)$$

where $m_1 = m_2 = \rho v_1 \omega_1 \Delta t = \rho v_2 \omega_2 \Delta t$; ω_1 and ω_2 are the cross-sectional areas of the pipe.

We obtain from (7.3): $\rho v^2 \omega (1 - \cos\beta) = R \sin\beta$; from which

$$R = \frac{\rho v_\omega^2 (1 - \cos \beta)}{\sin \beta} = \frac{\theta \rho v (1 - \cos \beta)}{\sin \beta};$$

where $\beta = 90°$

$$R = \sqrt{2} \, \rho v^2 \omega = \sqrt{2} \theta \rho v. \qquad (7.4)$$

The reaction force which develops in a pipe with a smooth bend during pulsations of flow for the output cross section of horizontal pipes can be determined by Bernoulli's equation

$$\frac{\rho v_K^2}{2} + p_K = \frac{\rho v^2}{2} + p. \qquad (7.5)$$

We obtain from (7.4) an expression for the reaction to a pulsation shock

$$\Delta R = R_{max} - R_{av} = \sqrt{2} \, \omega \, (\rho v^2)_{max} - (\rho v^2)_{av}. \qquad (7.6)$$

Or, with some simplification, assuming that the outflow speed is equal to 0, we obtain the maximum increase in reaction force acting on an elbow (greater than R_{av})

159

$$\Delta R = 2\sqrt{2}\, D p\, \pi r^2 = 8.855\, \Delta p r^2. \qquad (7.7)$$

Then the velocity of the mass of the pipe plus the fluid will be
$\frac{dx}{dt} = \omega A \cos \omega t$; the acceleration $\frac{d^2 x}{dt} = \omega^2 A \sin \omega t$. The force producing

this acceleration

$$P = -(\Delta m)\, \omega^2\, A \sin \omega t. \qquad (7.8)$$

This force is directed toward the position of rest.

The vibrations of a pipe and its effect on the pipe support are determined from velocity and acceleration of the oscillating motion (in this case, velocity is a derivative of displacement, while the acceleration is a derivative of velocity). The product of the vibrating mass of the pipe section times the acceleration is the force acting on the pipe.

If we consider that the vibration of an elementary mass Δm follows a sine wave, the displacement from the state of rest is

$$x = A \sin \omega t, \qquad (7.9)$$

where t is time; $\omega = 2\pi f$; A and f are the amplitude and frequency of the oscillations.

7.2. Fluctuations of Pressure in Vertical Strings

The pipe in water drainage, hydraulic hoisting and stowing systems in deep mines is subject to bending and bursting due to sudden pump stoppage and closure of the check valve on the discharge side of a hydraulic hoist pump [16]. For example, the maximum pressure reached

in the "Konrad-1" vertical pipe string was about 20.2 MPa, the minimum
pressure 7 MPa. The pressure depends on the speed of the flow before
stoppage (before sudden pump shutdown), the method of operation of the
protective device, the type of closure of the check valve, elasticity
of the pipe and fluid, moment of inertia of the pump impeller and its
electric motor. The test pressure in the pipe string reached 16 MPa,
the hydrostatic pressure reached 12 MPa, while the pressure in the pump
was almost double the normal operating pressure.

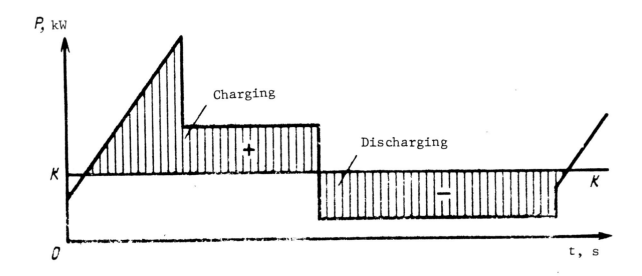

Figure 7.2. Graph of effectiveness of a flywheel.

The impact pressure decreases with a decrease in hydrostatic pressure
and an increase in distance from the focus of the impact in the direction
opposite to the flow of fluid. The drop in impact pressure is not
directly proportional to the distance from the pump to the midpoint of
the pipe.

The following measures were taken to protect the "Konrad-1" vertical
string from shocks: a flexural strength safety factor of 1.1 was used,
which was found to be insufficient and therefore the distance between
support was halved, increasing the strength safety factor to 4.5. The

pump drive was also given additional flywheel torque, by the attachment of an inertial mass, a steel casting which caused the pump to continue rotating after disconnection of the motor for an additional two minutes, given the impact phase t $\frac{2L}{a} = \frac{2 \cdot 600}{1000}$ = 1.2 < 2 min. The length used here is half the depth of the shaft, since the impact wave, according to our figures, is attenuated by the middle of the shaft. The flywheel completely eliminated the problem (figure 7.2).

The measures used for shock protection in the "Konrad-2" string are as follows: a safety valve, adjusted to the hydrostatic pressure in the pipe, is installed at the receiving suction valve of the centrifugal pump. Sudden disconnection of the pump may cause a separation in the column of fluid; therefore, the safety valve opens, emptying the pipe. The pressure fluctuates when this is done due to the resistance to the flow caused by the pump and safety valve.

Instead of an ordinary check valve, a wedge-shaped checking element was installed on the discharge side of the pump. This device has worked quite well in operation in automated drainage installations.

In case of a sudden shutdown of the pump, the valve closes the pipe in 5-30 seconds. The time of closure of the wedge-shaped valve is adjustable; when reverse flow starts, the valve is open and transmits the fluid through a large aperture, after which the aperture is smoothly closed.

The shock wave created by the wedge valve itself reaches its maximum shortly before complete closure, given the operating time of 10 s. The joint operation of the wedge valve and safety valve has decreased the maximum shock pressure to 7.5 MPa (minimum 2.5 MPa). After 70 s the pressure dropped to 4.8 MPa, and after an additional 300 s it reached the hydrostatic pressure of 4.5 MPa.

The test pressure in freely suspended pipes was taken as five times the hydrostatic and operating pressure at the bottom end, where the shock pressure was less than at the source of the shock (the pump), where it was 4.5 MPa.

Check valves rapidly lose their seal; therefore, a gate valve should be installed with them, and should have an independent hydraulic drive powered by a flow of water regulated by a differential piston. The pump must be selected with sufficient head reserve. When the pump is new, the excess pressure is choked to avoid overloading the drive. This is done by limiting the travel of hydraulically controlled gate valves, choking the flow as a function of the degree of pump wear. Considering the throughput capacity of the valve, the slide is adjusted to provide the required degree of choking. This causes velocity and wear to increase in the narrow section; therefore, an additional choke consisting of a permanent diaphragm is installed in the discharge pipe.

Pressure fluctuations in the pipe were tested by oscillographic studies, producing a typical diagram of an impact, beginning with a wave of pressure reduction. The check valve closed in one second, and the pressure rose from 3.9 to 6.6 MPa. Complete damping of the pressure occurred in 60 s. The speed of propagation of the hydraulic shock was 1000 m/s. It is recommended that the automatic control system shut off the fluid supply after first closing the gate valve on the discharge line to assure reliable operation and eliminate hydraulic shock.

Observations have established that the increase in length of a pipe upon hydraulic shock with sudden shutdown of the pump corresponds to the weight of the fluid in the string and the extension resulting from the sudden increase of pressure in the pipe. However, the lateral deflection of the pipe increases due to the pulsating pressure wave and its change in the direction of movement of the flow in the bottom elbow. Upon hydraulic shock, the pipe is shifted toward the center of the shaft. If the bottom end of the string is attached behind the elbow and the point of attachment is located at distance L from the elbow, as this distance increases.the stabilizing effect of the pipe mass G increases. At the maximum distance L (several meters), calculations should be performed using 0.25-0.5 G in place of the full value of G. Attachment of the pipe directly at the elbow is most undesirable, since the elbow cannot move and the system will be deformed. The "Konrad-2" pipe string was attached behind the elbow.

When filled with water the string oscillates slightly due to elongation caused by the internal pressure and an increase in overall weight, as well as the constant change in the direction of movement of the flow. The tensile force G, elongating and stabilizing the pipe, acts with some delay, which has never been calculated. The circular movement of the pipe stops after some time due to the nonuniform distribution of the weight of the pipe along its axis.

The parameters of a freely suspended self-supporting pipe and elbow should be selected using the maximum hydraulic shock pressure, which is then decreased by the protective measures indicated above. The string should not have guides or intermediate supports, which reduce its "play."

The vertical strings of deep mines, with anchor supports at the beginning and end, are equipped with temperature compensators: they are designed for the forces resulting from changes in pressure in the pipe and its deformation due to the changes in the air temperature in the shaft. When a gland compensator is used, the forces of friction in the guide collars P_h and gland compensator $P_{тp}$ and the forces from spreading of the compensator P_p under the influence of the pressure in the pipe on the compensator cup, i.e., on the circular cross section of the pipe [17] determine the force on the string

$$N = P_{н} + P_{тp} + P_{p}, \text{N.} \qquad (7.10)$$

Since the deviation of the line of the string from a straight line is slight, we can discard P_h; then

$$N = P_{тp} + P_{p}, \text{N.} \qquad (7.11)$$

The longitudinal deformation of the pipe causes forces which exceed the friction in the compensator, resulting in unloading of the stresses in the string. $P_{тp}$ and P_p in this case are transmitted to the top of the string from the resultant load (figure 7.3). From the compensator to the bottom support, guide collars are installed at certain intervals to give the string longitudinal stability. Design of the vertical pipe between spans is performed so as to achieve proper selection of the length of a span L and assure longitudinal stability and reliable operation of the string. The friction in the gland compensator

$$P_{TP} = \pi D_{CT} \, h f p,$$ (7.12)

where D_{ct} is the outside diameter of the cup; h is the height of the layer of packing in the gland, measured along the pipe; f is the coefficient of friction in the gland; p is the pressure in the pipe at the point where the compensator is installed, Pa.

The gland compensator spreading force

$$P_p = \frac{\pi (D_{CT}^2 - D_{BH}^2)}{4} p.$$ (7.13)

The critical Ritz force for a vertical string with simultaneous application of a uniformly distributed load and a concentrated load

$$P_{Kp} = P_E = \frac{Q}{2}$$ (7.14)

where P_E is the calculated critical Euler force; Q is the mass of the pipe considering the acceleration factor n = 1.2, from the gland compensator to the anchor support, ignoring the mass of the fluid, N.

The condition of stable operation of a string with a concentrated load in its upper portion is $Q < 2P_E$. When the rod spans between guide collars, the critical Euler force

$$P_E = \pi^2 \frac{E I_{min}}{l_\nu^2}$$ (7.15)

where E is the modulus of elasticity of the pipe metal, Pa; I_{min} is the moment of inertia considering the negative tolerance for the pipe wall thickness used, m^4; L is the length of a span between collars, m; ν is the strength safety factor, $\nu = 3.5$ for reliable operation. If the quality of manufacture of the string is low and the initial bending of the pipe is on the order of the wall thickness, then $\nu = 2$. The safety factor of a vertical pipe with eccentricity δ is $\nu = \nu_1$; $\nu = 2 \cdot 3.5 = 7.0$. Consequently,

$$P_E = \frac{1.41 \ E \ I_{min}}{l^2}, \qquad (7.16)$$

and from (7.14) considering (7.15) and (7.16) and replacing P_E and P_{kp} with force N we obtain

$$N = 1.41 \ \frac{E \ I_{min}}{l^2} - \frac{Q}{2}. \qquad (7.17)$$

The limit of applicability of the Euler formula for flexibility, the lower value of which $\lambda_{min} = \pi \sqrt{\dfrac{E}{\sigma_p}}$, if $\sigma = 2.02$ Pa, is $\lambda_{min} = 100$. The low flexibility of the system $\lambda < \lambda_{min}$ requires a low permissible load after Tetmajer

$$P = \frac{\omega - k}{\nu} \ (1 - a \lambda), \qquad (7.18)$$

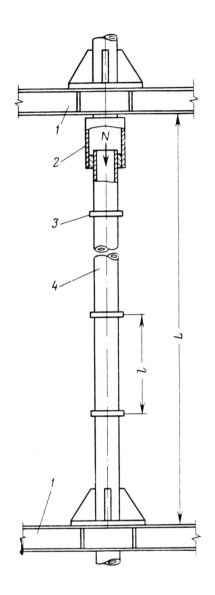

Figure 7.3. Diagram of attachment of a vertical string in a deep mine: 1—intermediate anchor support; 2—gland compensator; 3—guide collar; 4—pipe

where ω is the minimum cross-sectional area of the metal of the pipe considering the negative tolerance for wall thickness and wear, m^2; K = 3 for soft low-carbon steel; a = 0.00368 is an empirical coefficient.

166

For low string flexibility and deviations from linearity by δ, we use $\nu = 3.5-7$, in proportion to the rod flexibility λ.

The length of the pipe span between collars is determined by trial and error. The flexibility and strength safety factor ν are determined for various span lengths. The maximum permissible force P according to expression (7.8) is compared with the load $T = N + \frac{Q}{2}$ and the corresponding span length is selected [17].

7.3. Control of Vibration of Horizontal Pipes

Frequently, horizontal pipes are mounted on high supports and installed on the surface. The frequency of their forced oscillations f_b depends on the wind speed v, m/s perpendicular to the axis of the pipe, the pipe diameter D, cm, and the dimensionless Strouhal number S, which is 0.186 for circular pipe, i.e.

$$f_b = 0.185 \, \frac{v}{D} \text{ ,Hz} \tag{7.19}$$

In practice, the forced oscillations due to wind load do not exceed the permissible dynamic oscillation. However, if the frequency of the wind load falls within the range of the natural resonant frequency of the pipe, significant resonant phenomena may arise.

The oscillations due to wind load are calculated using an equation for determination of the oscillating frequency of the pipe

$$f_c = \frac{a}{l^2} \sqrt{\frac{EI}{Q}} \text{, Hz} \tag{7.20}$$

for rigid supports

$$f_c = 112 \sqrt{\frac{11000 \cdot 2 \cdot 10^6}{1518 \cdot 22^3 \cdot 10^6}} = 112 \sqrt{\frac{22}{15975}} = 4 \text{ Hz}$$

With articulated supports $a = 49.2^0$, the modulus of elasticity of the steel $E = 2.1 \cdot 10^6$ kgf/cm^2, the moment of inertia $I = 11,000$ kgf.

167

Pipe oscillations are controlled by the construction of vibration dampers, which have been successfully used for the 325 mm diameter pipe just described. The greater the mass m of a dynamic vibration damper, the less the displacement x_c. Design of a vibration damper must be based on the conditions required to stop the oscillations of the pipe

$$k_2 - m\omega^2 = 0 \quad \text{or} \quad \omega = \sqrt{\frac{k_2}{m}}. \qquad (7.21)$$

The amplitude of oscillations of the mass of a pipe section plus the vibration damper M and m should be $x_1 = 1$ and $x_2 = -\frac{P}{k_2}\sin\omega t$.

Standards for permissible oscillations of pipes. The permissible oscillations of foundations beneath devices with dynamic loads have been little studied. The experience which has been accumulated leads to the following analytic equation

$$2A = \frac{c}{\sqrt{n}}, \qquad (7.22)$$

where 2A is the amplitude of permissible oscillations, mm; c is a coefficient representing the quality of balancing (c = 1 is "excellent" balancing, c = 2 is "good" balancing, c = 3 is "satisfactory" balancing).

It has been established in practice that for machines with rotating frequencies of n < 400 rpm, the amplitude of oscillations should not exceed 0.19 for satisfactory balancing, 0.15 for good balancing, 0.1 mm for excellent balancing.

The standards for permissible oscillations were systematized by Robon, who drew up graphs of the variation of vibration as a function of speed of rotation recommended for powerful machines. With a rotating speed of 365 rpm (type 28Gr-8 soil pumps), the permissible amplitude of oscillations A = 0.2 mm, for n = 1450 rpm (type 3Gr-6 pumps) A = 0.1 mm, while for high speed coal pumps with n = 2800 rpm, the maximum permissible amplitude is only A ≤ 0.004 mm.

7.4. Designing Hydrotransport Systems for Hydraulic Shock

Hydraulic shock refers to a sudden change in the speed v or density ρ of a slurry, causing a rapid increase in pressure at the source of the shock, which then propagates in the direction opposite to the movement of the fluid. The result of a sudden change in v or ρ is, therefore, a sudden transition of the kinetic energy of the flow to potential energy at the shock pressure, such that the pressure in the pipe at the shock source instantly increases. Longitudinal failure of the pipe is an indication that the damage was caused by hydraulic shock (see figure 7.4).

The phenomenon of hydraulic shock during pipeline transport of grade G run-of-mine coal 25-55 mm in diameter was studied on a pilot installation with 169 mm diameter pipe, wall thickness $\delta = 11.5$ mm. The arrangement of cup-shaped strain gauge pressure sensors is shown in figure 7.5 [18]. The source of hydraulic shock was a plug cock, opened after 0.15 s, or a check valve, rotated horizontally 180° to block the moving flow modeling a sudden electric power failure. The coal slurry was fed through the installation by a 5ShNV coal pump. The first sensor S_1 was installed next to the check valve, sensors S_2 and S_3 at distances of 55 and 146.6 m from the first sensor (see figure 7.5). Figure 7.6 shows an oscillogram of the hydraulic shock. The shock wave

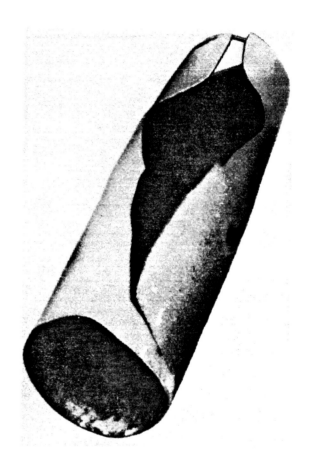

Figure 7.4. Failure of pipe resulting from hydraulic shock (D = 600 mm, δ = 10 mm)

169

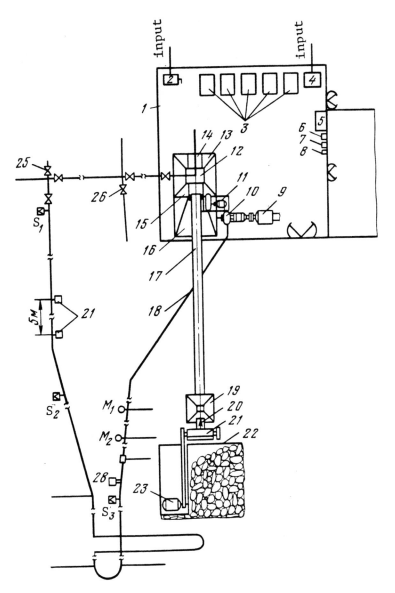

Figure 7.5. Diagram of pilot scale hydrotransport installation: 1--pumping plant; 2--380 V knife switch; 3--KF-22 resistor box racks; 4--type VYaP-6 high-voltage breaker; 5--three phase connectors for resistor leads from rotor circuit; 6--380V automatic starter for belt conveyor motors; 7-- same, for crusher motors; 8--buttons to connect and disconnect contacts (5) and starters (6,7); 9--high voltage asynchronous electric motor with phased rotor; 10--10U-4 coal pump; 11--belt conveyor motor; 12--graduated tank; 13--main sump; 14--rail bed; 15--sloping diaphragm in sump; 16-- secondary sump (for startup and washdown); 17--belt conveyor; 18--pipe (D = 160 mm); 19--container to receive crushed material; 20--sloping flume; 21--jaw crusher; 22--material storage area; 23--crusher motor; 24-- plexiglas inspection window; 25--pipe washout valve; 26--valve feeding slurry into longwall model; 27--differential manometer; 28--aircap.

170

traveled from the source of the disturbance in the flow to an aircap
with a volume of 1 m³. The pipe sections were joined by welding or
flange joints using pressboard as the gasket material. The pipe
sections were rigidly mounted to bracket supports in foundations.

The diagram shows significant attenuation of the leading edge
of the shock wave as it travels from sensors S_1 and S_2 (the head of
365-360 m water is reduced to 61 m water) to sensor S_3, 100 m distance.
The pressure rose primarily in the first zone to greater than the
initial pressure due to conversion of the kinetic energy of the
slurry to shock pressure, while a portion of the water continued
to move forward, compressing the free air present in the flow. The
solid phase (coal) continued to move after the pump was stopped due
to inertia and its initial velocity, acquired in the steady state.
However, after the liquid phase (water, small suspended particles)
stopped, the solid particles gradually lost their velocity. The lumps
of coal settled to the bottom with some delay after the moment of
formation of the shock pressure wave, the large lumps settling first,
followed by the smaller ones. The energy of the coal particles was
transmitted to the flow, increasing the shock pressure in the second
pressure increase zone. By the end of the second zone this process
had stopped, since there is sufficient time in longer pipes for settling
of the solid particles. The third characteristic zone for a shock is
an area of pressure maintained at the source of the disturbance for
time $t \leqslant \dfrac{2L}{a}$, (shock phase), after which the reverse wave relieves the
pressure.

It is recommended that hydraulic shock be calculated using the
following equation [18]:

$$\Delta H = a \ \frac{\eta}{g} \ v_{pac} \ [1 + k_2(nd - 1)],$$ (7.23)

where a is the propagation velocity of the hydraulic shock wave in
the slurry (for transportation of a three-component water-coal slurry);

171

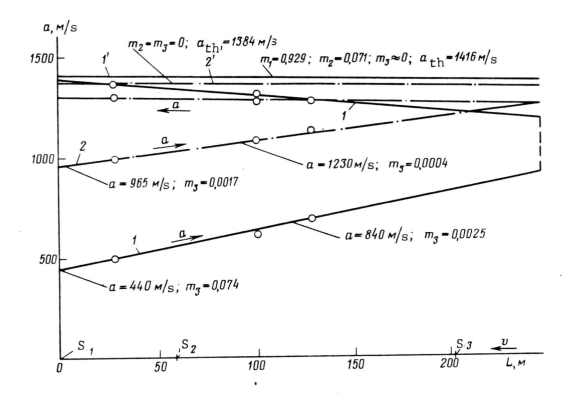

Figure 7.6. Hydraulic shock wave propagation velocity in a 150 mm
diameter slurry line: S_1, S_2, S_3 -- pressure sensors designed by the
author, arranged in the direction opposite to the direction of movement of the flow; 0 -- location of check valve at the end of the pipe;
1,2--change in velocity of slurry and water as a function of air
concentration in the flow; 1', 2' -- theoretical velocities.

$$a = \frac{a_0}{\sqrt{k_1\left[1 + \frac{(1-\mu^2)D}{E\delta}\varepsilon_1\right] + k_2\varepsilon_1\frac{\gamma_2}{\gamma_1}\left[\frac{1}{\varepsilon_2} + \frac{(1-\mu^2)D}{E\delta}\right] + k_3\frac{\gamma}{\gamma_1}\frac{R\varepsilon}{\Delta H}}} \qquad (7.24)$$

In calculations for steel pipes, Poisson's ratio μ = 0.28, the
modulus of elasticity E = $2\cdot10^6\,N/cm^2$. The volumetric modulus of
elasticity of water ε = 2100 N/cm^2, and then $\frac{(1 - \mu^2)\varepsilon_1}{E} \simeq 0.01$.
The propagation velocity of sound in water can be taken as a_0 = 1438 m/s
for pressures of up to 25 N/cm^2, water temperature 10°C; then equation
(7.24) becomes

172

$$\frac{a}{g} = \frac{a_0}{\sqrt{[k_1(1-0{,}01\frac{D}{\delta})+k_2\varepsilon_1\frac{\gamma_2}{\gamma_1}(\frac{1}{\varepsilon_1}+0{,}01\frac{D}{\delta})+k_3\frac{\gamma}{\gamma_1}\frac{Re_1}{pp_0}]}} \qquad (7.25)$$

where k_1, k_2, and k_3 are the concentrations of water, solids and air in
the slurry by volume; usually $k_3 = 0.001{-}0.006$, where $k_3 = 0.001$
$a = 969$ m/s, where $k_3 = 0.006$, $a = 611$ m/s; ε_2 is the volumetric
modulus of elasticity of the solid (coal); λ_1 and λ_2 are the specific
gravities of water and the solid material and the inside diameter and
thickness of the pipe wall [sic-tr.]; D/δ is the hydraulic shock
pressure considering the initial (preshock) pressure p_0; $\kappa = 1.41$ is
the adiabatic compression exponent $(1/\kappa = 0.171)$; v_{pacx} is the velocity
of the slurry before the hydraulic shock, m/s; $a = \dfrac{v_2}{v_1} = \dfrac{k_{2pacx}}{k_2}$; $n = \dfrac{\lambda_2}{\lambda_1}$

defined by the conditions of mutual slip between the solid and liquid
phases of the slurry [18]. In addition to the calculation equations
which we have presented, the specific conditions of separation of flow
from a pump upon sudden disconnection of the electric motor require
some refinement of the calculation equations as is the case where a
flow discontinuity develops (for example, at a sharp bend in the pipe).

Recording of the hydraulic shock with three pressure sensors
allows the settling of the solids as the shock pressure develops and
the transmission of energy to the flow to be traced (see figure 7.6
and 7.7).

The process of development of the shock wave is influenced by the
density of the slurry, particularly the presence of free undissolved
air in the liquid, in a concentration varying from 0.002 to 0.01. The
shock pressure characteristic with $\lambda_{coal} = 1.555 \cdot 10^4 \text{N/m}^3$ and $v = 3.42$ m/s is:

	S_1	S_2	S_3
Pressure sensors (see figures 7.5 and 7.6)			
Initial head H_0, m water	7.54	12.78	16.2
Shock value of head considering initial head, m water	365	360	61

We can see on the oscillogram that the pressure changes at sensor S_1 in 0.8 s (pressure rise zone), at sensor S_2 in 0.6 s, at sensor S_3 in 0.2 s. This example shows that the development of hydraulic shock should be prevented at a possible shock source (check valve, sharp bend in pipe in any plane, breaks in the line where continuity of movement may be disrupted) by installing protective devices.

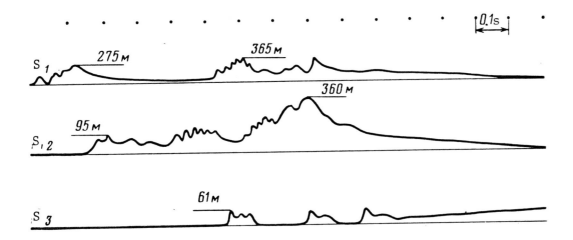

Figure 7.7. Oscillogram of hydraulic shock when coal is fed through 150 mm diameter pipe (diagram of placement of sensors S_1, S_2 and S_3 in figure 7.5). Distance between sensors S_1 – S_2 -- 55 m; S_2 – S_3 -- 143.6 m.

As the leading edge of the disturbance travels from sensor S_2 to sensor S_3 (see figure 7.6), a time of 0.4 s with distance between sensors L_{2-3} = 143.6 m, the propagation velocity of the hydraulic shock wave was found by direct measurement to be $Q = \frac{1}{t} = \frac{143.6}{0.4} = 359$ m/s, which indicates the presence of a significant quantity of free air in the flow. If there were no free air present, the propagation velocity of the shock would be twice as great. Figure 7.6 shows graphs of the change in the shock wave. The arrows indicate the

174

development of the forward and reverse pressure waves, pointing out
the significant increase in the force of the shock and its propagation
velocity (by a factor of 2) due to the compression of the mixture
(water plus air plus solid).

Cavitation of the fluid flow occurs under the following conditions:

$$p_0 - p_p < a \rho v_0; \quad (p_p = p_{pot} + p_{vac}).$$ (7.26)

Studies have shown that in pipes 50 to 1200 mm in diameter with
two- and three-component fluids moving through them, the general
nature of shock phenomena is identical and independent of the diameter
of the pipe. Studies have established that, in spite of the assumption
of N.Ye. Zhukovskiy that the greatest increase in pressure developing
along the length of a pipe should propagate at the speed of the shock
wave, the shock pressure varies along the length of the pipe, being
less, the greater the distance from the source of the shock. This is
because simultaneously with the increase in pressure in the pipe a
shock wave develops and propagates in the direction of flow of the
slurry. Even before the shock pressure reaches its maximum value in
the first phase of the shock process, this wave propagates throughout
the entire length of the pipe, is reflected from the end of the pipe
and begins to propagate in the direction opposite to the movement of
the flow (toward the source of the shock). It is this reverse wave
which prevents the propagation of the maximum pressure through the
entire length of the pipe. Therefore, countershock devices should
be installed in the immediate vicinity of a shock source (for example)
a check valve). Figure 7.8 presents graphs of the change in shock
pressure along a pipe, recorded at three points.

The decrease in velocity of a shock wave due to free air in the
slurry also decreases the shock pressure and the time required for
its attenuation; the duration of the shock phase is significantly
increased.

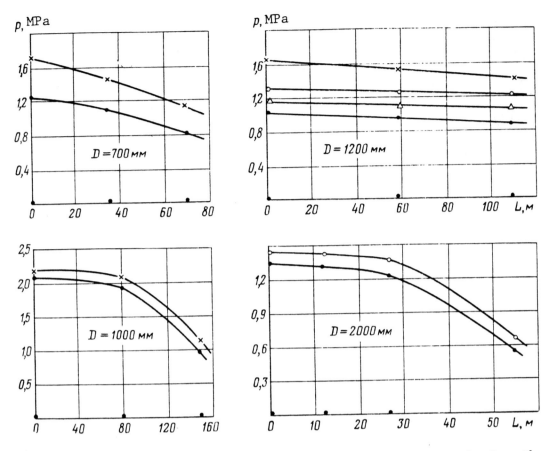

Figure 7.8. Graphs of change of shock pressure p along the length of a pipe for pipes of various diameters.

The concentration of solids has comparatively little influence on the propagation velocity of a shock wave where $k_2 < 0.3$; however, it significantly influences the nature of attenuation of the pressure oscillations, the duration of which decreases with an increase in concentration of the slurry [18]. Therefore, releasing air under atmospheric pressure or the operating pressure developed by a pump into the suction side of the pump is recommended as the most effective and least expensive means for damping hydraulic shock, particularly in pumps in the second and subsequent stages of a line, where p_0 may be greater than p_{atm}.

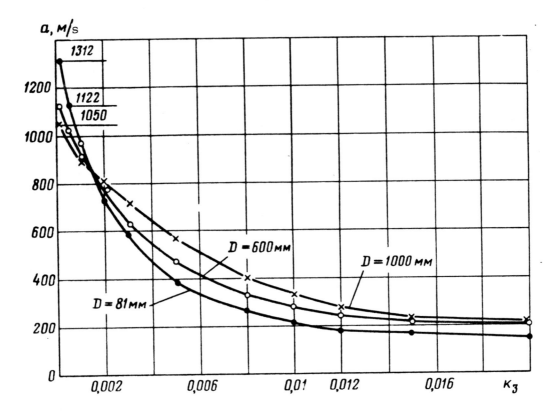

Figure 7.9. Graphs of the variation of a as a function of air concentration in the slurry k_3, constructed using calculation equation (7.27).

A simplification of equation (7.26) can be used to obtain an equation to determine the shock wave propagation velocity in a pipe carrying a water-air mixture, which can also be used for hydrotransport systems

$$a = \frac{a_0}{\sqrt{1 + \frac{L_{c_1}}{\delta E} + \frac{k_3 E}{p - p_0}}} \cdot \qquad (7.27)$$

Figure 7.9 shows theoretical graphs constructed using equation (7.27) and clearly demonstrating the influence of undissolved air in a slurry on the shock wave propagation velocity in a pipe.

Studies performed by the author have established the optimal quantity of air to be released into pipes of various diameters to prevent the development of hydraulic shock (table 7.1). The effectiveness of air is maximal as its concentration in the slurry is increased from 0.001 (ordinary slurry) to 0.01, decreasing the shock wave velocity from 1000 to 300 m/s, i.e., to less than one third. The increase in shock pressure over the operating pressure is reduced by the same factor.

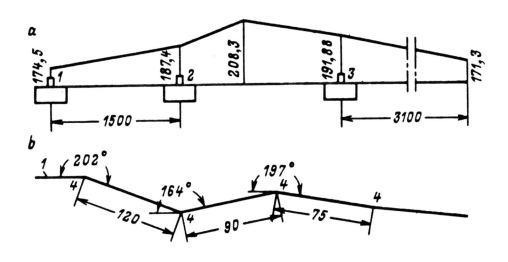

Figure 7.10. Possible hydraulic shock sources: a,b -- profile and plan; 1,2,3--pumping plants; 4-source of hydraulic shock.

Table 7.1

Pipe diameter D, mm	Speed of flow under stable conditions v_{sl}, m/s	Required concentration of air to damp hydraulic shock in medium being transported, %
50-300	1.2-2.5	0.25-1.5
400-800	2.5-4.0	0.45-1.5
800-1200	4.0-6.0	0.8 -1.5

Possible hydraulic shock sources are shown in figure 7.10 where sharp bends in the profile of the pipe create the danger of shocks. In case of a sudden disconnection of the coal supply, the continuity

178

is disrupted and the flow, moving in the direction opposite to the feed, collides with the now closed check valve or in the opposite direction collides with the separated mass of fluid, the force of the impact being proportional to the sum of the velocities of the fluid flows

$$\Delta p = a(v_1 + v_2)\rho.$$

(7.28)

A break in continuity of the flow may result from a sudden change in the profile of the line if a normally operating intake and exhaust valve are not present at the transition point.

When the continuity of a slurry flow is broken, the fluid continues to move at velocity

$$v_1 = v_0 - \frac{p_0 - p_v}{a\rho_0},$$

(7.29)

where v_0 is the initial velocity of the flow; p_0 is the absolute pressure in the pipe before the discontinuity arises; p_v is the absolute saturated vapor pressure of the fluid; a is the pressure wave propagation velocity; ρ_0 is the density of the slurry.

When the discontinuity cavity collapses, masses of fluid moving in opposite directions collide, leading to the development of hydraulic shock, which can be calculated by the equation

$$p - p_0 = \frac{a\rho_0 v_1}{\sqrt{1 + \frac{p_{req}}{p_1 - p_v + \rho_0 g H} \frac{v_1^2}{v_0^2}}},$$

(7.30)

where p is the absolute pressure in the pipe at the point of the flow discontinuity when the hydraulic shock occurs; v_1 and v_2 are the velocities of the oppositely directed flows in the cavity, m/s; p_1 is the difference in pressures at the ends of the pipe (where the slurry is dumped into the atmosphere $p_1 = p_{atm}$); g is the acceleration of the force of gravity; H is the slurry lift height.

179

Equation (7.30) can be used in horizontal and ascending lines with pipe slope of up to 30-40°. More studies are needed for steeply ascending pipe.

If after the flow discontinuity arises the pump is started before the reverse flow causes hydraulic shock, the oppositely directed masses of slurry collide. The shock pressure in this case is greater by the sum of the velocities of the colliding flows v_1 and v_2

$$\Delta H = a\,(v_1 + v_2)\,\rho.$$

The influence of pipe length on excess pressure upon hydraulic shock. The existing equations for determination of the variation in excess pressure upon hydraulic shock in a pressurized pipe include the hydraulic resistance coefficient, assumed constant over the length of the pipe. However, this assumption is incorrect, since λ (the coefficient of friction) changes with length during transient processes [19].

Theoretical analysis has resulted in the development of an equation for determination of the excess pressure of hydraulic shock considering the length of the pipe (restoration of the forces of friction is not considered). The equation does not include the quantity λ, variable over the length of the pipe, since its determination for unsteady processes is quite difficult [19].

Let us assume that there is a simple pipe with diameter V_0, length 1, through which a homogeneous slurry moves under steady conditions at speed v_0. If the end of the pipe is instantly closed, a forward hydraulic shock develops, the excess pressure in which above the steady state pressure is determined as follows according to the theory of N.Ye. Zhukovskiy:

$$\Delta p = \frac{a\,v\,\gamma}{g}, \qquad (7.31)$$

where a is the propagation velocity of the shock wave, m/s; λ is the specific gravity of the slurry, N/m^3; g is the acceleration of the force of gravity, m/s^2.

180

The hydraulic shock causes an increase in the diameter of the pipe
(figure 7.11), which can be determined by the equation

$$D = D_0 + \Delta D = D_0 + \frac{D_0^2 \Delta p}{2 \delta E_{\text{тp}}},$$ (7.32)

where ΔD is the pipe diameter increment; δ is the thickness of the
pipe wall; E_{tp} is the modulus of elasticity of the pipe, N/m^2.

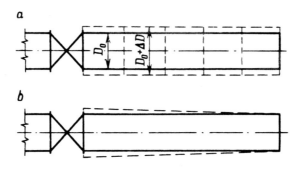

Figure 7.11. Diagram of development of hydraulic shock and increase
in pipe diameter.

To simplify our analysis of the process, we can assume that the
pipe diameter increases uniformly over a certain length (as is shown
in figure 7.11a). The rate of increase (change) in pipe diameter is
almost five times the speed of propagation of the hydraulic shock wave
through the slurry. If we assume that the pipe is divided into separate sections, the number of which is equal to n, the value of the
first section of length L./n, will be $\frac{L}{n} \frac{\pi D^2}{4}$, i.e., it will be greater

than in the steady state $\frac{L}{n} \frac{\pi D_0^2}{4}$.

The instantaneous change (increase) in volume causes a reduction in
the pressure over the section of pipe in question. This occurs due to

a decrease in the velocity within this volume; a negative pressure
wave is formed and propagates in the direction opposite to the movement of the flow. Let us determine the change in velocity on the
basis of the condition of continuity

$$v_0 \frac{\pi D_0^2}{4} \frac{L}{n} = v_1 \frac{\pi D^2}{4} \frac{L}{n},$$

(7.33)

where v_1 is the speed of movement of the flow after the increase in
pipe diameter, m/s.

From expression (7.33) we obtain

$$v_1 = v_0 \frac{D_0^2}{D^2}.$$

According to N.Ye. Zhukovskiy, a pressure pulse propagates in the
direction opposite to the movement of the fluid, its amplitude being

$$\Delta p_1 = - \frac{a(v_0 - v_1)\gamma}{g}.$$

As the hydraulic shock wave passes the plane of the end of the first
pipe section, the pressure change

$$\Delta p_1' = \frac{a v_0 \gamma}{g} - \frac{a(v_0 - v_1)\gamma}{g} = \frac{a v_1 \gamma}{g}.$$

After this the hydraulic shock wave reaches the second section, also
of length L/n. The same method can be used to determine for the second
section the same parameters which were determined for the first.

This analysis can be conducted for subsequent sections as well. Thus,
the increase in pressure at the end of the pipe as the hydraulic shock
wave passes

$$\Delta p_\ell = \frac{a v_0}{g} \gamma \left(1 + \frac{D_0}{2\delta} \frac{\Delta p}{E_{\text{тр}}} \right)^{2n}.$$

In equation (7.33), $n = f(L)$, which can be written as

$$n = \frac{kL}{2}$$

where k is a constant which is determined experimentally. The resulting equation can be written as

$$\Delta p_L = \frac{a\, v_0\, \gamma}{g}\left(1 - \frac{D_0}{2\delta}\ \frac{\Delta p}{E_{Tp}}\right)^{-kL}. \qquad (7.34)$$

Experiments were performed under both laboratory and industrial conditions to determine k, yielding the equation

$$k = \frac{1}{L}\ \frac{\ln \Delta p - \ln \Delta p_L}{\ln\left(1 + \dfrac{D_0}{2\delta}\ \dfrac{\Delta p}{E_{Tp}}\right)}.$$

The results of experimental studies are presented in table 7.2.

Table 7.2

Δp $9.8\cdot$ $10^4\ \frac{N}{M^2}$	Δp_1 $9.8\cdot$ $10^4\ \frac{N}{M^3}$	$D_0\cdot$ M	L, M	$\frac{\Delta p\cdot 10^6}{E_{Tp}}$	δ, M	k	k_{av}
32.05	28.10	0.081	170	152.6	0.00425	0.56	0.57
26.00	25.71	0.081	170	124	0.00425	0.58	
38.25	37.74	0.150	200	184	0.0110	0.57	
33.16	32.78	0.150	200	158	0.0110	0.56	0.57
30.47	30.13	0.150	200	145	0.0110	0.59	
24.92	24.70	0.150	200	119	0.0110	0.56	
16.80	13.90	1.200	110	79.9	0.0120	0.51	
13.40	11.90	1.200	110	63.8	0.0120	0.54	0.55
11.70	10.2	1.200	110	55.0	0.0120	0.55	
10.70	88.0	1.200	110	50.9	0.0120	0.59	

As we can see from the table, k changes over a very small range for a rather broad range of experimental conditions, allowing us to consider our theoretical analysis correct.

Similar results were obtained by comparison of the theoretical calculations presented above with the experimental data of other researchers.

Equation (7.34) can be written as

$$k = \frac{1}{L} \cdot \frac{\ln \frac{a\, v_0}{g\, \Delta p_L} \gamma}{\ln\left(1 + \frac{D_0}{2\delta} \frac{\Delta p}{E_{Tp}}\right)}.$$

If we assume that $\dfrac{a v_0 \lambda}{g \Delta p_1} = e$ (where e is Napier's number), the

equation above can be written for e as

$$e = \frac{1}{k \ln\left(1 + \frac{D_0}{2\delta} \frac{\Delta p}{E_{Tp}}\right)}. \tag{7.35}$$

The physical sense of equation (7.35) is as follows: the inverse value of coefficient K divided by $\ln\left\{1 + \frac{D_0}{2\delta} \frac{\Delta p}{E_{tp}}\right\}$ yields the length of

pipe over which the amplitude of the excess pressure during a hydraulic shock decreases by a factor of "e".

The conditions of development of a hydraulic shock in two-layer or lined pipe are special, complicating the flow of multicomponent fluids through these pipes [20].

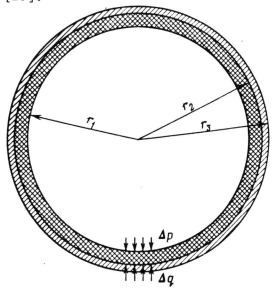

Figure 7.12. Diagram of two-layer pipe lined with cast basalt.

Let us agree that the layer in contact with the fluid will be called the inner layer. We can represent (figure 7.12) the radii of the inner layer as r_1 and r_2, the outer radius of the pipe as r_3. Suppose E_1, μ_1 and E_2, μ_2 are the modulus of elasticity and Poisson's ratio for the inner and outer layers of the pipe, respectively.

N.Ye. Zhukovskiy derived the equation

$$\Delta v = a \left(\frac{\Delta \gamma}{\gamma} + \frac{\Delta f}{f} \right), \tag{7.36}$$

where Δv is the absolute instantaneous change in velocity of the flow caused by a sharp disturbance of the flow, m/s; $\Delta\gamma/\gamma$ is the relative change in weight of the fluid per unit volume; $\Delta f/f$ is the relative increase in cross-sectional area of the free space in the pipe; a is the hydraulic shock wave propagation velocity, m/s.

Partial coverage of the cross section of the pipe (for example, by a gate valve) causes an increase in the interior pressure by Δp. The outer layer of the pipe prevents expansion of the inner layer, causing reverse pressure Δq, directed inwardly along the radius of the pipe, to develop at the surface of contact between the two layers.

According to (7.35) we have

$$\Delta q = \Delta p \frac{1 - A}{\left(\frac{r_2}{r_1} \right)^2 - A}, \tag{7.37}$$

where A is the elastic characteristic factor;

$$A = \frac{0.01 \, E_1 - (1 + \mu_1) \, n}{0.01 \, E_1 + (1 - \mu_1)(1 - 2 \, \mu_1) a};$$

n is the specific resistance factor, determined by the expression

$$n = \frac{0.01 \, E_2 \left[\left(\frac{r_3}{r_2} \right)^2 - 1 \right]}{(1 - \mu_2) + (1 + \mu_2) \left(\frac{r_3}{r_2} \right)^2}.$$

185

Representing

$$B = \frac{1 - A}{\left(\dfrac{r_2}{r_1}\right)^2 - A},$$

expression (7.37) becomes

$$\Delta q = B \Delta p. \tag{7.38}$$

Let us study the plane deformation of the cross section of the inner layer under the influence of the internal pressure Δp and external pressure Δq.

With an accuracy equivalent to small second order quantities we can write

$$\frac{\Delta f}{f} = \frac{2 \Delta r_1}{r_1}.$$

Applying the equation of Lame to this case and substituting expression (7.38) into it, we obtain

$$\frac{\Delta f}{f} = \frac{2 \Delta p}{E_1} \left(\frac{r_2^2 + r_1^2}{r_2^2 - r_1^2} + \mu_1 - \frac{2 B r_2^2}{r_2^2 - r_1^2} \right). \tag{7.39}$$

As we know from the theory of elasticity

$$\frac{\Delta \gamma}{\gamma} = \frac{\Delta p}{\epsilon}, \tag{7.40}$$

where ϵ is the volumetric elasticity modulus of the medium which, in the case of a two-phase flow $\epsilon = \left(\dfrac{k_1}{\epsilon_1} + \dfrac{k_2}{\epsilon_2} \right)^{-1}$; k_1; ϵ_1; k_2, ϵ_2 are the volumetric concentrations and elasticity moduli of the water and the solids, respectively.

Therefore, (7.40) becomes

$$\frac{\Delta \gamma}{\gamma} = \left(\frac{k_1}{\epsilon_1} + \frac{k_2}{\epsilon_2} \right) \Delta p. \tag{7.41}$$

We know from the theory of hydraulic shock that

$$\Delta v = \frac{g \Delta p}{\gamma a} \tag{7.42}$$

where g is the acceleration of the force of gravity; γ is the weight per unit volume of the slurry; $\gamma = k_1 \gamma_1 + k_2 \gamma_2$; γ_1; γ_2 are the weight per unit volume of the water and the solids, respectively.

Substituting expressions (7.39), (7.41) and (7.42) into (7.36), after a few transformations we obtain

$$a = \frac{1}{\sqrt{\dfrac{\gamma}{g} \left[\dfrac{k_1}{\epsilon_1} + \dfrac{k_2}{\epsilon_2} + \dfrac{2}{E_1} \left(\dfrac{r_2^2 + r_1^2}{r_2^2 - r_1^2} + \mu_1 - \dfrac{2 B r_2^2}{r_2^2 - r_1^2} \right) \right]}} \tag{7.43}$$

Using information obtained from practice, changing the thickness of the lining within the limits used (for example, basalt castings), and also varying the pipe diameter, we used (7.43) to calculate a, the propagation velocity of a hydraulic shock wave (table 7.3), which shows the velocity both with and without air in the flow. These data were used to construct graphs (figure 7.13) of the variation of a as a function of lining thickness δ and pipe diameter D.

Table 7.3

Pipe diameter D, mm	$k_3 = 0,007$				$k_3 = 0$			
	Lining thickness δ, MM				Lining thickness δ, MM			
	20	30	40	50	20	30	40	50
500	474	476	477	478	1212	1241	1262	1278
800	469	472	474	475	1129	1168	1197	1220
1000	466	469	471	473	1083	1126	1159	1185
1200	462	466	468	470	1041	1088	1124	1153

The equation did not consider the presence of air in the flow. The equation for a single-layer pipe carrying a three-component flow (water plus solid plus air) is already known [18]

$$a = \frac{a_0}{\sqrt{k_1 \left(1 + N \frac{D}{\delta}\right) + \frac{\gamma}{\gamma_2} - k_3 \, c \, \frac{\delta_1}{\Delta H_i}}},$$

where $N = \frac{(1 - \mu^2)\varepsilon_1}{E}$; E is the modulus of elasticity of the pipe; $c = 1 - \left(\frac{H_0}{H_0 + \Delta H_i}\right)^{1/\kappa}$; $\kappa = 1.41$ is the adiabatic compression exponent of the slurry in the process of hydraulic shock.

Then for a three-component flow equation (7.43) is written as follows [18, 21]

$$a = \frac{1}{\sqrt{\frac{\gamma}{g} \left[\frac{k_1}{\epsilon_1} + \frac{k_2}{\epsilon_2} + \lambda k_3 - \frac{2}{E_1} \left(\frac{r_2^2 + r_1^2}{r_2^2 - r_1^2} + \mu - \frac{2 B r_2^2}{r_2^2 - r_1^2}\right)\right]}},$$ (7.44)

$$\lambda = \frac{1 - \left(\frac{p_{aT} + p_0}{p_{aT} + p}\right)^{1/\kappa}}{p - p_0},$$

where p_0 is the initial pressure; p is the hydraulic shock pressure at the moment of determination of the shock wave propagation velocity; p_{at} is atmospheric pressure.

As the results derived demonstrate, air in this case decreases the hydraulic shock wave propagation velocity to less than half, and also decreases the magnitude of the shock.

From the standpoint of pipe wear resistance, it is desirable to increase the thickness of the lining, which also increases the rigidity of the system. Equation (7.43) allows us to select an effective pipe diameter and liner thickness. For example, the shock wave propagation velocity in a pipe 800 mm in diameter with a 20 mm thick liner is almost equal (see table 7.4) to the shock wave propagation velocity in a 1200 mm diameter pipe with a liner thickness

188

of 40 mm, meaning that the use of the 1200 mm diameter pipe with liner 40 mm thick is more effective.

Thus, in spite of the increased rigidity of the pipe system and the increase in danger of hydraulic shock, this quantity does not increase greatly when a cast stone liner is used.

Means for protection from hydraulic shock widely used in low pressure installations (heads up to 100 m water), consisting of double or single aircaps, are used in dredges and the tailings areas of mine beneficiation plants. For higher pressure hydrotransport systems (transporting coal, ore, concentrates, sand and gravel) and water lines, a system of bypass lines for repumping plants with steeply ascending and broken profiles has been suggested (figure 7.14).

As we can see from figure 7.14, the bypass line has a dual-seat check valve, the design of which is shown in figure 7.15. As the multi-stage system, including the pumps at the repumping plant, is started up, the flow of slurry, bypassing the pump, opens the dual-seat check valve, the disc of which in this case covers the bypass line. In case of a sudden electric power failure, the check valve allows the slurry to move into the bypass line around the pumping plant in the opposite direction, preventing hydraulic shock [21].

The shortcomings presently seen in the planning of equipment and selection of multi-stage slurry feed systems make it impossible to achieve the full effectiveness of this very convenient and inexpensive type of transport. For example, in selecting slurry pumps and pipes, their hydroabrasive wear and the influence of wear on the operating conditions of the installation depending on the properties of the medium being transported are not considered, causing systems with flow cavitation to be selected for multi-stage transportation of slurries.

Analysis of the hydroabrasive wear of the impeller of a centrifugal pump has shown that when abrasive fluids are transported, pumps should operate at lower speeds with larger impeller diameters. Consequently, in selecting pumps for operation in hydroabrasive media, low-speed machines should be considered, and when high-throughput, high-head

189

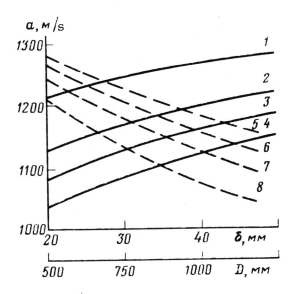

Figure 7.13. Variation in hydraulic shock wave propagation velocity a as a function of thickness δ and diameter D of lined pipe: 1-4--a as a function of δ with D = const; 5-8--a as a function of D with δ = const

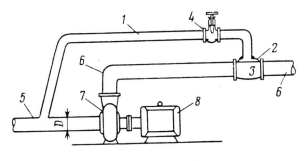

Figure 7.14. Diagram of pumping plant bypass: 1--bypass line; 2--two-seat check valve; 3--position of valve during operation of pump; 4--gate valve on bypass line; 5-6--main slurry line; 7-8--coal pump and electric motor

slurry pumps are designed, the diameter of the impeller should be increased rather than the operating speed. If the solid material being transported allows it, the number of vanes on the impeller should be increased, which increases the wear-resistance of the impeller and helps to improve the efficiency of the pump.

Considering the results of hydro-abrasive wear, it is better to transport slurries of high concentration. This decreases the specific wear of the vanes and increases the quantity of solids which can be fed through the impeller.

Studies have shown that for multi-stage transport of slurries, the pumps should be connected in series and operate without flow separation. This is achieved by complete automation and proper operation of the system, also increasing the throughput of solids while significantly decreasing capital investment.

Studies on the protection of the pressurized pipes of hydrotransport systems from hydraulic shock have shown that the specifics of operation (presence of solids, frequent startup and shutdown) require the development of new designs.

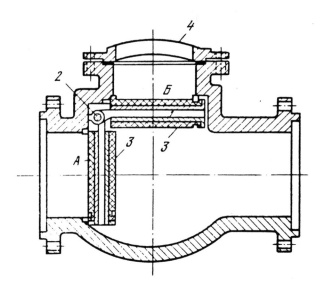

Figure 7.15. Two-seat check valve: 1--steel disc; 2--shaft; 3--seal
(rubber, capron); 4--bypass pipe seat; A--position of valve during
operation of bypass; V--position of valve during normal transportation of slurry through main line.

Statistical data on the development of hydraulic shock during
the operation of pressurized hydrotransport systems and the loss
resulting from this phenomenon can be used to analyze the need for
protection of similar systems from sharp pressure fluctuations, as
is called for by the technical conditions and planning standards
(TU and NP) for systems for hydraulic shock control in the pressurized
pipes of hydrotranport systems developed at the Institute of Mining
Mechanics, Georgian Academy of Sciences.

The TU and NP call for the determination of: the propagation
velocity of the hydraulic shock wave in pipes during transport of
the slurry and the maximum excess pressure above the steady state
pressure; the necessary number of pumps for multi-stage transportation

191

and the head in the suction pipes of the pump required for normal operation when they are series-connected.

Studies of the specifics of operation of hydrotransport installations are under way at the Institute of Mining Mechanics. Designs for hydraulic shock protective devices have been developed, as well as methods which can be effectively used under the conditions in question. Protective devices with bypass lines have been developed for multi-stage systems. The most successful designs have been investigated and used under production conditions, and also given to planning organizations for use in the planning of pressurized hydrotransport systems.

To protect high-head pumping plants used in the first stage of such systems and locations where hydraulic shock can be expected to develop, the use of caps filled with rubber motor vehicle inner tubes, hoses or balls to reinforce the elastic energy damping effect (figure 7.16) is recommended. These designs, installed in the pumping plants of a single-stage slurry system using 700 mm diameter pipe and a lift height of 85 m, have yielded positive results. Hydraulic shocks and their undesirable aftereffects upon sudden disconnection of the electric power have been eliminated [21]. The cap is installed beyond the pumping plant immediately after the check valve. To increase the effectiveness of the elastic elements (hoses or balls), air must be pumped into them to generate excess pressure. The hoses can be laid out in rings. The end of each hose should be bent and tied tightly.

Inner tubes are the best quality soft rubber dampers which can be used in these caps. They can be inserted in rings by selecting the cap diameter (given the required volume) 10-15% larger than the diameter of the tubes. Damaged tubes can also be used, by cutting chunks out of them and holding them in place with an ordinary metal clamping plate and a few bolts. The pieces of inner tube can be lowered vertically into the cap. To prevent them from being sucked into the pipeline, a small screen with aperture smaller than the

192

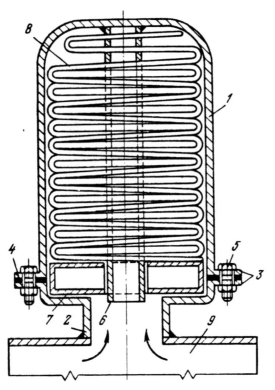

Figure 7.16. Diagram of aircap filled with rubber hoses (balls or
inner tubes): 1--body of air column; 2--connecting fitting; 3--flanges;
4--gasket; 5--M-20 bolt; 6--guide rod; 7--hollow disc; 8--12 mm diameter
hose; 9--main line pipe.

inner tubes (or balls) should be placed across the input.

In case a powerful shock develops (collision of oppositely moving
flows due to separation following sudden loss of power or sudden
restoration of power), a portion of the fluid is drained through a
fast acting safety valve as shown in figure 7.17.

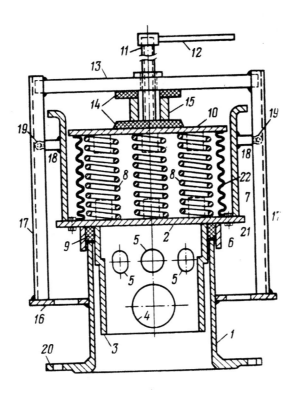

Figure 7.17. Fast acting safety valve: 1--body; 2--cover; 3-4-5--slurry
dumping aperture; 6-7--metal cylinders; 8--spring; 9--elastic seal;
10--metal plate to transmit tightening force; 11--rod; 12--tightness
regulation arm; 13--beam; 14--elastic shock absorber; 15--sleeve;
16-20--flange; 17--channel sections; 18--spacer bar; 19--rollers;
21--bolts; 22--bellows

CHAPTER 8   OPERATION OF HYDROTRANSPORT SYSTEMS

8.1.   Reliability of Operation of Hydrotransport Systems

Control of the condition of equipment and mechanisms is an important part of the process of control, particularly of such continuous
systems as hydraulic mining and coal transport systems. The elements
of the equation for P, profitability, determine the relationship
between the status of the equipment and the man-machine system [22]

$$P = \frac{P_b - \theta}{C_{o.c.} + C_{HO.C}},\tag{8.1}$$

where $P_b = P_d - C_{ct}$ is the profit; θ represents deductions from
profit; C_{oc} is the cost of invested capital, including equipment;
$C_{HO.C}$ is the cost of materials and spare parts (standardized operating
capital); P_{ts} is the selling price; $C_{ct} = \frac{z}{\theta_p}$ is the cost; Z represents
expenses related to production of the product; θ_p is the quantity of
the product produced.

Assurance of operability of machines, fittings and pipes, maintenance and repair are several times more costly than new machines,
primarily due to hydroabrasive wear. Automation of control increases
these costs still more, but assures high productivity and continuous
operation of systems over long periods of time. The high cost of
repair and maintenance of equipment results from insufficient information on the actual condition of the equipment. The unavailability
of diagnostic procedures to prevent failures and the lack of proper
maintenance and care mean that only defects which are made obvious
are repaired, increasing the time and funds spent for this purpose [22].

Planned down time required by the standard operating procedures
may amount to up to 30% of the total working time. The operation of

a hydrotransport system with good quality equipment and pipes plus reliable information on the condition of the equipment, achieved by the use of testing and adjustment points, sensors and test stands allow redundant hydrotransport systems (with one operating line and one backup line for repair) to be avoided.

The increase in the length of transport and in the capacity of hydrotransport systems used under difficult climatic and geographic conditions require greatly increased reliability. The USA has set up a "Department of Pipeline Reliability," which studies and systematizes pipeline emergencies, developing technical conditions for their construction, repair and operation, performs inspections to monitor the fulfillment of the standards which have been developed and studies the possibility of using plastic pipes for the transport of toxic and corrosive media. This department also develops new law designed to strengthen the process of monitoring the construction and operation of pipeline systems, assuring the observation of standards for welding under field conditions and cathode protection of pipes.

The operation of pipeline systems has shown that the conditions of operation of pipelines are degraded by poor operation or complete failure of gland compensators. Pipes frequently sag due to settling of the soil and sinking of supports, pipe joint welds and flange joints are sometimes poorly made, or fail due to changes in stresses in the pipes. Pipes frequently corrode on the outside or inside due to poor quality paint or the effects of the products being transported on the pipe metal. Hydraulic shock also moves the pipe longitudinally (according to our data, where D = 1000 mm, ρ = 1.15 t/m^3, p = 1.5 MPa, with a pressure rise Δp = 2 MPa the pipe may move by 5 to 20 cm).

Changes in operating conditions also change the frequency of vibrations from 5-7 to 12-14 Hz during startup and shutdown of pipelines, causing the amplitude of oscillations to increase by a factor of 1.5-2. The natural flexural oscillating frequency can be estimated as the bending of a beam, and approaches the frequency of shock and emergency conditions.

Resonant phenomena in the case of sudden shutdown of pumps and separation of the flow causes nonuniform stresses (greatest at the center of spans), which are increased by ovalness of pipes and rapid hydroabrasive wear.

Recently, the modernization of operating equipment and introduction of new equipment have increased the reliability and improved the operating characteristics of machines and mechanisms, helping to improve the operation of hydraulic mines.

The operating conditions of coal pumps determine their service life. Processing of statistical data, with 8-34 rotors in a sample, variation factor -.13-0.3, indicates that: the operating life of coal pump rotors is proportional to the suction generated by the coal pump (figure 8.1) and inversely proportional to the length of the suction line (figure 8.2). As the water temperature rises from 15-29°C, the operating life of the rotor increases by 18-20% as a result of increased evolution of gas. Operating life reaches its maximum at a certain head, as we can see from the figures below.

Head range, KPa	Statistical mean operating life rotor, hr	Life of rotor, %
19.6-0	296	45
98-196	655	100
588-833	495	74
980-1372	325	48

Identical, maximum service life of all machines in a hydrotransport station is achieved where their suction lines have low, identical resistance, by proper selection of pipe diameters [24].

The nature of failures in the operation of monitors, based on the results of testing of the 12GP-1 monitor, indicates that of the nine causes of failure, five result from poor quality manufacture of parts or welding (shaft pin and hydraulic cylinder angle fitting, hydraulic panel mounting plate). Failures up to 124, 185, 360, 425, and 500 hours of operation result from wear of the following parts: seal due

197

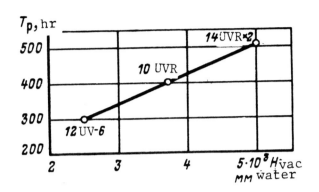
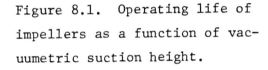

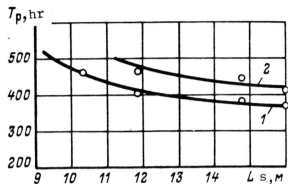

Figure 8.1. Operating life of impellers as a function of vacuumetric suction height.

Figure 8.2. Variation in operating life of coal pump impeller as a function of suction line length: 1-10--UVR x 2; 2--14 UVR x 2

to damage to chrome-plated surface, bucket turbine vanes, replaceable bushing in support and interchangeable bushing in spider, interchangeable spider bushing.

8.2. Pipeline Operating Experience

The most characteristic type of damage to pipes in hydraulic mines is a defect in the sealing devices and valves, extrusion of rubber sealing rings (gate valves have a mean time between failures of only 1.5 months). Rubber sealing rings most frequently fail due to poor installation. It has been found that if a ring is replaced, it fails due to flange wear. Ring failures occur at one point around the circumference; it is therefore desirable to rotate pipes by 90-180° relative to each other. Hydraulic shock has been observed in water lines, resulting from switching of the high pressure water supply from face to face. Hundreds of cases of pump shutdown due to excessive pressure have been recorded in hydraulic mines.

Typical pipe emergencies are caused by erosion, particuarly at bends.

It is very important to measure the wall thickness of pipe periodically without stopping the transportation process, and to test pipe specimens for strength periodically in order to assure reliable operation.

A method is being developed for repair and sealing of underground pipe to eliminate leaks. In order to provide access to the damaged section a small diameter bore hole is driven and dirt is removed around the pipe section. A sealing solution is then fed in the hole to the damaged section, completely covering it. To prevent leakage of the product, a pressure higher than the pressure in the pipe is created in the bore hole. Sealers used include tars, asphalt, cement and other materials.

The Japanese have studied the reasons for the formation of plugs during operation of pipelines. Three groups of reasons of formation of plugs have been considered:

inclusion of particle size larger than the maximum permissible size in the slurry;

a change in particle size distribution and viscosity of the slurry;

a decrease in the speed of movement of the slurry.

As the viscosity of the slurry increases, it takes on the properties of an anomalous fluid, and the initial shear stress in the slurry increases. A device has been patented in Japan for maintenance of the optimal transportation speed, which monitors the thickness of the sediment. This is done by installing sensors in the pipe at predetermined intervals to check the viscosity and speed of the moving slurry. The signals of the sensors are transmitted to a central monitoring point by radio. The information from the sensors is delivered to an electric motor, the speed of the motor is changed, thus maintaining the speed of movement of the slurry constant.

Failure of pipes due to soil corrosion or wear, when it results from a change in pressure (startup, hydraulic shock, formation of a plug) usually occurs along the pipe. Most failures are less than 500 mm

199

in length (or 75%), and can be repaired by placing sleeves or collars around the pipe [18].

The characteristics of the technology of hydraulic mining and hydrotransport of coal in hydraulic mines determine the nature of the quantitative and qualititative characteristics of the consumption of materials and equipment. The quantity of materials of the equipment consumed is one-half to two-thirds as great as in mines with the ordinary technology, 19-28 tons per 1000 tons of coal won. However, the labor consumption of transport of coal is twice as high (2.8-4.0 man-shifts per ton as opposed to 0.9-2.1 man-shifts per ton). This is because electric locomotives are used for the transportation of other cargos and personnel in ordinary mines. As we improve the operations of hydraulic mines, we must decrease the quantity of supplementary cargo carried. At "Yubileynaya" mine of "Kuzbassugol'" Union, the quantity of supplementary transportation is 10.8 tons per 1000 tons of coal won by hydraulic breaking with anchor supports.

The movement of cargo and personnel from the surface for the area next to the vertical shaft to the face is continuous, without changing the mode of transport, using a suspended monorail which can travel through openings with slopes of up to 35° and can operate in openings 2-2.2 m wide regardless of the properties or condition of the surrounding rock and soil, whether flumes and pipes are present or not. Monorail tracks are inexpensive to install and remove.

The monorail system in the Yaroslavskiy mine ("Kuzbassugol'" Union), which operates with three DMV-5A deisel locomotives, has proven reliable in operation and achieves a savings of 90,000 rubles per year.

8.3. Materials for Pipes

At the present time, steel pipe is usually used for the hydrotransport of bulk materials. This pipe is convenient to use, easy to install and remove, but has comparatively short service life and is quite expensive. Pipe with a wall thickness of 8-12 mm is usually used for coal mines in hydraulic mines. The service life, with preventive rotation of the pipe around its longitudinal axis, averages four years.

The service life of pipes can be increased by increasing the wall thickness. Pipes with wall thickness of up to 40 mm are used for this purpose abroad. Steel pipe can be replaced with ceramic, basalt and other pipes. The relative mean annual cost of operation of a slurry line is proportional to the wall thickness.

Calculations have shown that an increase in the wall thickness of slurry lines 273-325 mm in diameter from 10 to 15 mm results in an increase in service life by a factor of 1.8 and a decrease in the mean annual cost of operations by 23-29%; increasing the wall thickness to 40 mm increases the service life by a factor of 6, and decreases the mean annual cost of operation by a factor of up to 2.6.

Obviously, the use of pipe with greater wall thickness for slurry lines is economically justified.

It is expedient to increase the strength of pipe by the use of high-chrome alloys when slurry speeds are low, 2-3 m/s, i.e., for 150-200 mm diameter pipe. Larger diameter pipe requires high slurry speeds and hydroabrasive wear resistance is poor if this pipe is protected with high-chrome alloys. Silicon-manganese steels are more suitable for large diameter pipe. These steels take on increased wear resistance due to hardening resulting from the impacts of solid particles moving at high speeds.

Pipe lined with basalt has good wear resistance. At speeds of 2-5-10 m/s, the wear resistance of this pipe is 11.0, 7.7 and 5.9 times greater than the wear resistance of pipe made of type 4 steel. However, basalt-lined pipe is heavy, though no heavier than thick wall steel pipe, and is sensitive to impact loads. The question of selecting the material for pipe must be answered considering both technical and economic factors. There are no universal materials desirable in all cases.

Bimetallic pipes are of significant interest for the hydrotransport of abrasive materials. UkrNIIGidrougol' has tested two varieties of bimetallic pipe, 219 and 325 mm in diameter, with the thickness of both outside and inside layers in the 219 mm diameter pipe 5.7 mm, in the 325 mm diameter pipe 5.9 mm. The tests showed that the wear resistance of bimetallic pipe is three to six times greater than the wear resistance of pipe made of st.3 steel.

Cast stone is the most commonly used of nonmetallic materials, and is particularly suitable for the transportation of such highly abrasive materials as metallurgical slag or fossil fueled electric power plant slag. Plywood pipe is also usable.

Basalt inserts are pipe sections up to 500 mm in length with wall thickness 20 mm. The service life of a pipe with basalt inserts is several times greater than that of steel pipe. They can be successfully used in main transportation lines.

Multi-layer plywood pipe 300 mm in diameter, wall thickness 13 mm, has been tested at 1.5 MPa. Wear occurs only at the bottom of the pipe. Pipe joints are made using plywood collars. The wear is proportional to the time of exposure of the pipe walls to the slurry and depends on the concentration of solid particles. The wear is most rapid at joints, particularly if the two sections are not coaxial. Rapid wear is observed at bends.

Titanium Enterprises (Florida) uses rubber pipe for the hydrotransport of abrasive materials. The pipe diameter is 400 mm, and the flexibility of the rubber pipe allows the dredge to move without shutting down operations. The rubber pipe is designed for a service life of ten years, as opposed to two years for steel pipe.

Metal pipe lined with polyurethane is manufactured in Great Britain. As studies have shown, pipe with wear-resistant linings have a service life 8-20 times higher than that of steel pipe; it is intended for the hydrotransport of highly abrasive fine slurries. Depending on the abrasiveness of the slurries, the thickness of the wear-resistant lining may vary. The pipe diameter is 76 mm.

In the USA, Abrasist is used. This material is very strong, durable and capable of withstanding abrasive wear. Abrasist has excellent wear qualities. In West Germany, after 20 years of operation of this type of pipe transporting ash, no significant wear was noted.

The Japanese have patended a method of manufacturing pipe of flexible sheet materials. The pipe can be used for the transportation of coal, sand and other slurries. Pipe with inner linings of abrasion-resistant plastics is also used. The plastic is applied hot in three layers. Glass-fiber pipe intended for use under pressures of up to 1 MPa is manufactured in West Germany.

Glass pipe has been used extensively in the coal lines of mine beneficiation plants in the USSR. This pipe is particularly effective when surrounded by a thin wall steel shell, especially for areas with corrosive water (Donets basin).

8.4. Theoretical Prerequisites of Wear

Hydroabrasive wear, according to the latest research, results when abrasive particles impact against a surface. During the first phase of hydroabrasive wear, repeated impacts against the surface cause plastic deformation of the metal, and the metal is work-hardened. However, the hardening process continues only until the plastic properties of the metal have been exhausted. As the impacts of the solid particles continue, the degree of work-hardening increases, and the metal begins to lose its plastic properties. The upper layer cracks and flakes off as the abrasive particles continue to impact it.

In the second phase, particles of the metal flake off in one or more neighboring areas. The cycle is then repeated for the "fresh" surface thus exposed. It has been found that the rate of wear (K) which refers to the rate of loss of mass of a specimen, depends on the speed of the abrasive stream

$$K = av^m,$$

(8.2)

where K is the degree of wear corresponding to 1 kg of abrasive;
a is a coefficient which depends on the properties of the material
and corresponds to the intensity of wear at a flow speed of 1 m/s;
m is the exponent, which depends on the properties of the material.
According to experimental studies m = 2.5-3.2.

The main parameters influencing the wear rate are the speed of the
slurry and the angle of attack of the abrasive particles against the
pipe surface. The nature of the influence of the solid abrasive on
hydroabrasive wear of steel is determined by the ratio of hardnesses
of the worn material H_m and of the abrasive H_a.

If $K = \dfrac{H_m}{H_a} \leqslant 0.6 - 0.8$, this equation is linear. Where K > 0.6-0.8,
a rapid increase in the wear resistance of the materials with increasing
hardness is observed, one of the signs of the transition from direct
fracture of the surface layer to multi-cycle types of wear.

In some cases, rubber coatings are quite effective, since they
increase service life in comparison to the operation of steel pipe.
Rubber coatings are quite useful for finely ground materials, less
suitable for large lump materials.

The physical and mechanical properties of a slurry depend on the
particle size, shape and particle size distribution of a solid material,
its density and many other factors. The great variety of slurries prevents unambiguous determination of their abrasiveness.

The conditions of movement of the slurry (speed of movement of the
flow and distribution of solid particles along the length and over the
cross section) are quite significant. It is important that the concentration of solid particles in a slurry be practically uniform in
various cross sections of the pipe. The mean spatial concentration
may not give an accurate concept of the nature or magnitude of wear.

Wear of pipes is complicated also by the fact that the influence
of corrosion is frequently great. It is not always easy to distinguish
corrosive wear from erosive, if the results of the wear are considered.

The criterion of wear of pipes and equipment for hydrotransport must consider not only technical, but also economic factors. Wear can be decreased for given values of throughput of an installation, as slurry and as solids, and given physical and mechanical properties of the solid, by decreasing the speed of hydrotransport. However, this involves the danger of plugging of the pipe, i.e., the speed may drop below the critical speed. Also, reducing the speed requires an increase in capital investment, due to the increased pipe diameter required.

Even for homogeneous slurries, decreasing the speed of hydrotransport requires an increase in capital investment. Therefore, the problem is an ordinary one of optimization of the economic pipeline diameter considering the wear factor.

The abrasiveness of a slurry can be determined most accurately by considering the following factors:

1) the speed must be such that the conditions for selection of the optimal parameters of the hydrotransport system are satisfied for the specific objects at hand;

2) the particle size distribution, absolute particle size and other properties of the solid material must be as they are in the actual pipeline. Particular attention must be given to see that the material is not rounded, i.e., that its abrasive properties are not artificially reduced.

Since closed circulating systems are almost always used under laboratory conditions, in which the material passes through the transportation equipment repeatedly, it is impossible to achieve complete similarity of hydroabrasive to actual usage conditions. In determining the wear of a pipe, feeding apparatus can be used and the wear of the pipe separated from the wear of the other equipment, though this is a difficult task. A circular pipe also fails to solve the problem, since the same batch of material moves continuously and its abrasive properties are constantly decreased as its moves through the test loop.

The diameter of the pipe and the material of which it is made should be as in nature. The question of the influence of pipe diameter or pump size on pipe wall wear has not been well studied.

205

Thus, the abrasiveness of a slurry, determined by the quantity of material which must pass through the hydrotransport system to wear away 1 mm of wall thickness can be determined only under natural conditions.

For comparative analysis the abrasiveness of a slurry can be determined by the same methods used in machine building to determine the service life of machines on test stands. Severe conditions with rapid wear are set up on the test stand. Then, after performing a large number of experiments and comparing the results with reports on the operation of the equipment under natural conditions, a conversion factor is found allowing laboratory test results to be converted to operational conditions.

Therefore, the processing of data from many experiments is of decisive significance. Once reliable data are available on the wear of hydraulic equipment under natural conditions and test stand data are also available, we can make the transition to prediction of the wear of hydraulic equipment. It is assumed that the wear is proportional to the time of exposure to the abrasive medium. This assumption is not always accurate, since the abrasiveness of particles is decreased as the particles travel through pipes; equipment, particularly pumps, are constantly exposed to new, unrounded particles. If the degree of wear is determined in a laboratory test stand with the material passing through the test stand only once, the data can be converted more reliably. Existing methods of determination of wear using radioactive substances are difficult, dangerous and unreliable. Direct measurements of the wear of pipe walls by weighing of specimens or measurement of the wear of equipment using ultrasonic thickness meters, now in series production in our country, provides results which are closer to nature.

The operating principle and design of laboratory test stands may vary, and they are all not suitable for use as analogs for natural installations. Information on the abrasiveness of slurries must be obtained under natural conditions.

8.5. Wear of Coal Pumps

Modern pumps for the transfer of slurry are manufactured of the following materials, depending on the particle size and abrasiveness of the solid particles: manganese steel alloys for the transportation of large particles; martensitic "Nihard" -type cast iron for medium-sized particles; high-chrome cast iron and rubber for small particles.

Extensive operating experience has shown that rubber coatings are particularly suitable for low-head pumps, producing up to 40 m head, with particle sizes 3-10 mm.

The selection of a material should consider economic factors, not only wear resistance but also cost.

Parts of equipment which are exposed to hydroabrasive wear must operate under difficult conditions; wear depends on many factors: the speed of the slurry, the angle at which the slurry flow strikes the part in question, the concentration of solid particles in the slurry, the properties of the solid particles, such as size, degree of round-ness and hardness.

Wear depends to a great extent on the material of the part. The steels used are rapidly worn and must be frequently replaced. Their service life under production conditions can be increased by various methods: heat treatment; surface hardening (by high frequency induction or oxyacetylene flame); surface chemical and heat treatment; electrolytic coating -- chrome plating; the application of coatings of enamel or rubber; the use of materials with high hydroabrasive wear resistance, etc. Highly alloyed cast iron and steel have been found to be most wear resistant. For example, boron cast iron is 11 times more wear resistant than st.3 steel, high-chrome cast iron type SS is 4.9 times more wear resistant. Highly alloyed cast iron and steel are, however, not always suitable for pump parts. These metals are difficult to cut, weld poorly, tend to form cracks, are brittle and operate poorly with impact loading. Parts made of these alloys cannot be repaired after local wear by surfacing, which greatly limits the use of cast iron and steel alloys for coal pump parts.

The most effective means of increasing the wear resistance of coal pump parts exposed to hydroabrasive wear is to surface them with wear-resistant alloys. This method of strengthening of parts is most realistic when performed in mine equipment repair plants, since it allows the service life of parts to be increased several times over.

There are several methods of surfacing of the operating surfaces of parts: automatic surfacing with powder wire; semi-automatic surfacing with plate electrodes; manual surfacing with an electric arc using powder mixture; manual surfacing with oxyacetylene flame.

8.6. Wear of Pipes

The Skochinskiy Mining Institute has performed studies to determine the mechanism of hydroabrasive wear. Studies using piezoelectric sensors and rapid kinescopic surveying have shown that the abrasive particles act on the wall of the pipe by impacting the wall with all modes of movement of the solid particles through the pipe.

Various pipe sections differ in the nature of wear; at the head end of the pipe, particles of the hardened surface are chipped off. As the angle of attack decreases the process of chipping plays a less significant role, and abrasion becomes more significant.

When corrosion is present the process of hydrobrasive wear increases greatly. This is explained by the fact that thin oxide films are formed on the surface of the metal, which are mechanically weaker than the metal itself. These oxide films are worn away and then the fresh metal surface is oxidized again. This accelerates the process of wear.

Crushing of the material transported occurs simultaneously with wear of the pipe. Wear of the pipe and crushing of the material have the same nature. To minimize pipe wear, the speed of movement of the slurry must be minimal. The wear is significantly influenced by the quantity of rock which is present in the slurry. Elimination of rock not only helps to decrease the wear of the pipe, but also reduces the degree of attrition, since rock particles cause rapid wear of coal particles.

Studies have shown that pipe wear is a very complex process. Various factors such as the impact effect of the particles and their abrasiveness may wear the pipe nonuniformly, forming scratches and ridges of significant depth and height on the pipe wall, causing local wall weakening and increasing the hydraulic drag.

Hydroabrasive wear is influenced by a number of factors which are determined by the parameters of the pipe: its diameter, condition of the walls, material of the pipe, condition of joints, as well as the parameters of the slurry: properties of the solid particles, their density, shape, presence of projections, microscopic cracks, concentration and, finally, the conditions of movement of the slurry, i.e., its speed.

The durability of the pipe frequently has a decisive influence on the economic effectiveness of hydrotransport, since the cost of replacement of pipe makes up 20-50% of the entire cost of operation of a slurry installation.

An important factor influencing pipe wear is the speed of movement of the slurry, which must be selected considering both technical and economic factors. The recommended speed may be significantly greater than the critical speed, and must be computed considering the rate of crushing of the solid particles.

In the mines of the Kuznets basin, with slurry consistencies of 1:20-1:30, the throughput capacity of pipes as solids varies from 0.6 to 1.2 million tons of coal. Increasing the consistency to 1:6 decreases the specific wear of pipes by 30%, although the absolute pipe wall wear rate may increase; the throughput capacity of the pipes as solids also increases to 2.0-2.6 million tons.

Studies have shown that the diameter of a pipe has little influence on the specific wear of the pipe (the same phenomenon is true of the crushing of the coal which occurs during transportation).

The wear of a pipe results from two factors -- chemical corrosion and mechanical wear or erosion. When finely ground coal is hydraulically transported, corrosive wear is most important, the presence of oxygen in the slurry having some influence on this by causing oxidation of the metal walls of the pipe.

209

The degree of corrosive wear can be decreased by the use of additives which react with the atmospheric oxygen dissolved in the slurry.

Experiments have been performed to determine the effect of additives: whereas without additives pipe wall wear reached 2.9 mm per year, after sodium was added to the pipe, wear decreased to 0.038 mm per year. The optimal quantity of additive was then determined, decreasing the annual wear of the pipe to 0.02 mm, i.e., decreasing corrosive wear by a factor of more than 140.

Erosive wear can be decreased by grinding the material to be transported and decreasing the speed of movement of the slurry. The wear of pipes increases exponentially as a function of the speed of movement of the slurry. The exponent is 2.1-2.9, depending on the properties of the material transported and the pipe walls.

Wear is significantly influenced by the particle size of the material transported. This influence is insignificant once the particle size is reduced to 0.07 mm. As the consistency of the slurry increases, absolute wear also increases; however, the wear per ton of transported solid material decreases.

The walls of pipes are worn nonuniformly, the severest wear occurring on the bottom of the pipe, the least at the top. The walls of horizontal pipe are worn more nonuniformly than the walls of sloping pipe. The higher the speed of movement of the slurry above the critical speed, the less the nonuniformity of pipe wall wear around the perimeter. Hydroabrasive wear of pipe walls causes them to be polished smooth, although the opposite phenomenon is also observed, the walls becoming rougher under the influence of abrasive particles.

Increasing the wear resistance of pipes is not only a technical problem, but an economic problem as well, since it requires the use of more expensive materials, which may be justified only within certain limits. The optimal cost of a pipe must be determined as a function of the service life and cost indicators, for both capital investment and operating cost, per unit of product transported.

Studies have shown that grinding of the material in the process of hydrotransport is closely related to hydroabrasive wear of pipeline equipment.

Usually, before a slurry is transported, the pipe is filled for 30-45 minutes with water. After this, the pipe transports slurry for 17-20 hours; before shutdown, the pipe is washed out with process water again for 80-1000 minutes. The hydrotransport system is then shut down and the pipe walls contact air. Thus, the walls of the pipe are in contact with the oxygen of the air for 4-7 hours in each cycle, resulting in the formation of an oxide film. During this time, a layer of oxidized metal, a fraction of a micron thick, is formed. During subsequent startup of the hydrotransport system this film is worn away, then a new film is formed, etc. The loss of the film upon startup of the system greatly intensifies the process of corrosion of the pipe.

Pipes are exposed to the combined influence of erosion and corrosion. The pipe walls are worn away, and cracks are formed, the dimensions of which correspond to the thickness of the zone of increased wear.

8.7. Recommendations for Calculation of Probable Wear of Hydraulic Equipment

In order to prepare recommendations for prediction of the wear of pipes and other hydraulic equipment as a part of planning of long range hydrotransport systems, UkrNIIGidrougol' has undertaken special experimental studies and analyzed information available on research performed both in the USSR and abroad on this problem.

Recommendations for pipe wear in long distance hydrotransport must be based on the results of field studies under industrial conditions.

Let us analyze problems of pipe wear under natural conditions of long term operation. We know that the service life of water pipe is as great as 30 years and that hydrotransport of ash after combustion of coal is performed on a broad scale at coal-fired electric power plants. Special tests were undertaken at UkrNIIGidrougol' to compare the abrasiveness of ash and coal.

211

Studies performed at the Donets basin regional electric power plant showed that the service life of St.3 and St.4 steel pipe without rotation of the pipe was two years, or four to six years if the pipe was rotated three or four times. Slag and ash are transported through pipes up to 400 mm in diameter, concentration 3-4% by weight. No precise data on the quantity of material transported is available, though approximate calculations show that 12,500 tons of material can be transported for each millimeter of pipe wear. The hydrotransport of gangue in shaft mines yields approximately the same results.

Comparative tests show that the rate of wear produced by power plant ash is four times the rate of wear produced by coal fines with a maximum perticle size of 0.6 mm.

Considering the results produced, we find that the design service life of pipes for main hydrotransport lines, ignoring corrosion control measures, will be at least 10-13 years.

If we consider the results of study of the influence of inhibitors on wear, it becomes impossible to predict the service life of pipes using laboratory studies. This work must be done under field conditions.

The most complete data on pipe wear in long distance hydrotransport of finely ground coal have been accumulated in the USA. The wear of a 173 km pipeline in Ohio was found to be one-tenth to one-twentieth the anticipated wear based on test results produced by American researchers, once more confirming the correctness of the opinions of those researchers who predicted that the wear of equipment used in long distance hydrotransport cannot be modeled. Very large particle coal was transported, containing 35% volatiles, up to 9% ash, internal moisture content 1.8%, sulphur content 2.5-3.5%, particle diameter 0-2.4 mm, consistency S:L = 58:42, content of free hydrogen ions in the slurry P_H = 7.5. The inside diameter of the main pipeline is 254 mm, the speed of transportation 1.37-1.98 m/s, throughput 250-360 m^3/hr. The installation operated 8500 hours per year (utilization factor 97%), annual throughput 1.06-1.83 million tons. The material of the pipe was steel, tensile strength 300 MPa. Numerous studies of pipe sections revealed no cases of significant wear.

The particle size distribution of the solid material at the beginning and the end of the pipe was almost identical. Microscopic analysis showed that the grains of coal mostly had sharp edges, and little rounding occurred during transportation through the pipe. The American specialists believe that the pipe is worn primarily due to corrosion, and erosive wear is practically insignificant. The erosive wear of the pipes is not directly related to the quantity of coal transported. Neutralization of the corrosiveness of the medium and other measures have reduced the corrosive wear of pipes to not over 0.02-0.09 mm/yr, which is apparently not the ultimate limit.

Sodium bicromate and hexometaphosphate are successfully used to control corrosion in the USA (1.4 weight parts of sodium bicromate per 100 weight parts of slurry, with an equal portion of sodium hexometaphosphate). When inhibitors are used, the following wear rate is observed in the pipe (table 8.1).

According to approximate data, the cost of protection of a pipe from corrosion is 1.6 Kopeks per ton.

The service life of a pipe in a newly planned hydrotransport installation for finely ground coal with protection from external corrosion should be considered to be on the order of 20 years.

Various methods of testing wall thicknesses can be used under production conditions.

The ultrasonic method of wall thickness measurement is most widely used in hydraulic mines. The type USP-M-10 or "Kvarts-6" ultrasonic pulse thickness meter is used. Pipe wall thickness can be measured at practically any point on the pipe without shutting down the hydrotransport system.

The measurements produced are used to construct a pipe wear map through the cross section of the pipe and along the length of the line, allowing decisions to be made concerning the need to replace the pipe or rotate it around its longitudinal axis.

The abrasive properties of coal, which determine its capability to cause wear at any given rate, can be conveniently estimated in practice,

Table 8.1

Content of chromium ions	Increase in inside diameter, mm			
	beginning of pipe	at first repumping plant	at second repumping plant	at end of pipe
Planned	0.44	0.31	0.26	0.21
14 parts chromium per mm parts	0.02	0.04	0.09	0.09
25 parts chromium per mm parts	0.04	0.03	0.07	0.08

by values of hardness. A distinction should be made between aggregate hardness and the hardness of the rock-forming minerals. It is the latter which describes the abrasiveness of the material, whereas the aggregate hardness determines the "degree of consolidation" of the abrasive particles.

The microscopic hardness method is used at present to determine the hardness of coal and rock. Table 8.2 shows the values of microhardness of the components of the main types of coal from the Kuznets basin.

This table allows us to select an analog for determination of the presumed wear intensity during transport of other materials with no physical and mechanical properties.

Considering the data of table 8.2, we can determine the throughput capacity of the pipe during transport of coal of known physical and mechanical characteristics. We shall use the results of industrial observation of hydroabrasive wear of the slurry lines of the coal mines of the Donets basin, Kuznets basin, etc. (data from VNIIGidrougl', UkrNIIGidrougol').

The throughput capacity of planned pipes (thousands of tons) to the point of complete wear can be determined by the calculation equation

$$Q_2 = Q_1 \, k_d k_{cx} k_v k_D k_\delta \, {}^n \psi,$$

$$(8.3)$$

214

where Q_1 is the throughput capacity per mm of wear of the pipe wall; k_d is a coefficient considering the particle size of the coal transported; k_c is a coefficient considering the concentration of the slurry; $k_v = \left(\dfrac{v_{an}}{v_{pl}}\right)^2$ is a coefficient considering the speed of hydrotransport

(v_{pl} is the planned speed of transport of the slurry; v_{an} is the speed of transport of the slurry in the pipe-analog); k_D is a coefficient considering the diameter of the pipe (with $D_{av} = 200\text{-}400$ mm, we can assume $k_D = 1$); $k_\delta = \dfrac{\delta_{pl}}{\delta_a}$ is a coefficient considering the wall thickness of the planned pipe and the pipe-analog); n is the number of operating positions of the pipe (usually n = 1 to 4); ψ is the pipe wall wear nonuniformity factor.

Table 8.2

Coal type	Yield of volatiles	Microhardness of coal components, N/mm²		
		Fusain and zylosain	Vitrainized material	Yellow bodies
B	55	77	27	38
BD	50	119	32	41
DG	42	232	42	85
G	40	.152	37	86
Zh	35	420	48	119
KZh	26	434	45	-
K	23	485	52	-
PS	17	406	56	-
T	8	405	123	-
A	4	422	214	-

The values of Q_1, k_d, k_c, n and ψ can be determined by observation of the hydroabrasive wear of pipes under production conditions.

Example. Initial data for calculation: $Q_{sl} = 1200$ m³/hr; D - 400 mm; $\delta = 12$ mm; v = 2.55 m/s; s = 8% (by volume); $\rho_t = 1.45$ t/m³. Type G coal, 0-100 mm in diameter, is transported.

215

We know that when type G coal 0-100 mm in diameter is transported under the conditions of the hydraulic mines in the Kuznets basin at speeds near the critical speed, a 350 mm diameter pipe can transmit approximately 200,000 tons of coal per mm of wear of the bottom wall. Assuming $k_d = k_S = k_D = 1$, we find that

$$Q_\Sigma = 200\ 000 \cdot 1 \cdot 1 \cdot 1 \cdot 1,8 \cdot \frac{9}{1} \left(\frac{2,5}{2,65}\right)^2 = 2,900,000\ \text{м}^3.$$

For round-the-clock operation of the hydrotransport system 300 days per year, a pipe would fail after

$$T = \frac{2\ 900\ 000}{24 \cdot 300 \cdot 96 \cdot 1.45} = 3\ \text{years},$$

where $Q_T = sQ_{av} = 0.08 \cdot 1200 = 96\ \text{m}^3/\text{hr}$ is the hourly throughput of the system as solids.

The wear of pipes in hydraulic mines according to VNIIGidrougl', where 38 to $232 \cdot 10^3$ tons of rock mass is transported per mm of wall thickness, occurs primarily in the area of maximum wear and depends on many factors. Rotation of pipes two times by 120° can triple the service life of slurry-transport pipes.

The durability of coal pumps depends on their design, the material of which they are made, the abrasiveness of the slurry transported and the operating conditions. A number of methods have been developed to increase the durability of coal pump parts: mechanical and electrical hardening, heat treatment, chemical and heat treatment, the use of plastics, silicates and other coatings.

The French Fapkro Company produces pump equipment for use in coal mines and beneficiation plants. The pump bodies are made of white cast iron which has Brinell hardness of 450-500. At the points where the coal impacts the walls, the walls are thickened to 40 mm and, furthermore, equipped with antiwear armor discs which have a hardness of 600 units on the same scale; impellers have a hardness of over 500 units. Bearings have slipping casings and, as they wear and the sound of the pump changes (due to the increased clearance) they can be tightened with two adjustment bolts, thus retaining the initial centering and characteristics of the pump until the impeller wears out almost completely. The casing also protects the bearings from water and dust. The bearings are lubricated with grease or mineral oil.

Wear resistant lining chutes. Feeding of an abrasive and corrosive mixture consisting of 30% coal less than 13 mm in diameter and 70% water containing 5000 mg/l sulphuric acid through steel flumes made of St.3 steel in the mines of Kizelugol' resulted in a wear rate of 55.6 $g/m^2 \cdot hr$, while type 1Kh 13 steel was worn away at 1.29 $g/m^2 \cdot hr$, i.e., the wear resistance of the latter type of steel was four times greater. Steel type 1 Kh 18N9 has the same wear resistance. In the Donets basin with v_{sl} = 0.63 m/s a slurry consisting of coal with particle diameter 10-15 mm wears flumes of St.3 steel 12 times more rapidly than flumes made of 4Kh 13 chrome steel.

In the hydraulic mines of the Kuznets basin and Donets basin, recycled water is primarily mildly alkaline (pH = 6-8.3), the content of carbon dioxide reaching 118 mg/l only at "Krasnogorskaya" mine. Therefore, at these hydraulic mines corrosion occurs due to the influence of oxygen on the surface.

It is quite important to note that the nonuniformity of wear of flumes is increased by the presence of steps 70-100 mm high (quantity 30-75%). Transverse wear occurs on the longitudinal axis and decreases in the direction of the sides of the flume. In sloping flumes (60-70% of those in mines) the wear center is displaced in the direction of the slope.

Experimental data obtained by a number of research institutes and production facilities confirm the following characteristics of wear resistant materials carrying coal slurries.

1) at "Yubileynaya" mine administration, rubber 9 mm thick and a conveyor belt 10 mm thick were mounted in a flume using epoxy glue; the rubber lasted four months, the conveyor belt two months. Rubber lining is useful in chutes and steep cross cuts.

217

2) VNISKh has found that St.2 steel has 13 times greater wear resistance than glass fiber, while caprolin is 4.3 times more wear resistant, polystyrene is 6.8 times more wear resistant and polyamide is 3.25 times more wear resistant. VNIIPTUuglemash lined glass-reinforced plastic flumes with wear-resistant steel, producing flumes 30-40% lighter and 7-8 times more durable than ordinary steel flumes. The linings were made of "Makhroti" stainless steel 0.8 mm thick, and also of 1Kh 13 steel, more readily available.

Cast basalt is 7-8 times more resistant to corrosive and mechanical wear than St.3 steel. However, attachment of basalt and slag-sitall plates with cement stone paste did not yield good results (UkrNIIGidrougol'). When type 400 cement was used, the flume weighed over 300 kg; the use of screen reinforcement made the flumes still heavier (the screen was attached by hooks); they were therefore made in several parts, with the screens held in place by bolts in rubber gaskets.

VNIIGidrougol' surfaced the bottom of flumes to produce longitudinal waves 1.5-2 mm thick with PP-AN-170 powder wire beneath a layer of flux, with the waves spaced 10 mm apart. This increased the cost of the flume by 21%, while increasing the service life by a factor of 4.5 in comparison to flumes of ordinary St.3 steel. Flumes installed in the southern cross cut of "Yubileynaya" mine administration's hydraulic mine number 2 were still quite usable after passage of 380,000 tons of coal, with no visible signs of wear, as were basalt-lined flumes (held in with cement and epoxy resin). Type "G" coal had been transported, ash content 16-20%, particle size 0-250 mm, flow speed 3.5-4.5 m/s, slope of cross cut 0.005-0.007. Flumes with welded longitudinal ridges are recommended by VNIIGidrougl' for main and section openings.

In recent years, cast stone products of diabase, basalt and other rock materials have been widely used and have shown good results where there is no dynamic or high specific pressure on the wear surface. Lining plates, pipes, hydrocyclones and other products are made of silicate melts. When manufacturing cast stone products, a number of recommendations must be considered to maximize their effectiveness.

In the process of solidification and in the solid state, silicate melts have low heat conductivity, with significant linear and volumetric shrinkage, high hardness, low tensile and flexural strength; products of these materials tend to form cracks. Products of cast stone must meet requirements of durability and high reliability in operation, simplicity and economy of manufacture. High quality castings are obtained by the use of a technology of simultaneous or directed solidification of the parts of the casting. The thickness of the products must assure the necessary reliability, durability, rigidity, as well as ease of manufacture of quality castings. Very thin walls may develop cracks or fissures or may be short-run. Overly thick walls increase the mass of the products and facilitate the formation of shrinkage and gas pores. Plates measuring over 500 mm and pipe 300 mm or more in diameter are reinforced with metal frames made of round rolled products. The wall thickness T of a product is determined by the equation

$$T = 2 (A + B), \text{ мм,}$$

(8.4)

where A is the diameter of the rolled product for the frame, mm; B is the distance from the frame to the outside surface of the casting wall, mm, which is 1.2 to 1.5 times the diameter of the rolled metal.

The tolerances for cast stone products are 50-80% greater than for steel products, which is required by technological factors to assure good quality installation of the cast products, held in place by mortar, acid-resistant cement or a rock paste developed by V.G. Kazenyy. Pipes used in the hydraulic mines of the Kuznets basin transport up to 170,000 tons of coal per mm of wall wear or, if rotated four times, Q_g = 625,000-1,200,000 tons of coal. The minimum wall thickness in the bottom portion of the wall is 3 mm.

The cost of 1 km of pipe is 37,950 rubles; the cost of the coating in this pipe is 0.12·37,950 = 4250 rubles, or 4.25 rubles per meter. The cost of enamel coating of one meter pipe at the Krasnokamskiy pipe plant is 0.84 rubles. We therefore have a cost reserve. Sitall or special wear-resistant enamel coatings are two to four times stronger.

The coating should provide impact resistance and hardness. At speeds of not over 5 m/s, according to S.P. Kozyrev, cavitation wear can be ignored.

The microhardness of the most hydroabrasive component at the solid point of quartz is H_a = 1200 kgf/mm^2. The microhardness of pipe coatings H_p must be at least 0.6 H_a ≤ H_m or H_m = 0.6·1200 = 720 kgf/mm^2 ≅ 1000 kgf/mm^2. Enamel coatings have H_m = 800 kgf/mm^2. Crystallized sitall coatings have microhardness H_m = 1200-1300 kgf/mm^2, while metal ceramic coatings made from suspensions have H_m = 1700 kgf/mm^2. Ordinary enamel coatings satisfy the requirements of hardness, but are simpler in technology of application than sitall and metal ceramic coatings.

With a maximum lump weight of 3 kgf (radius of spherical lump R = 8 cm with specific gravity of material ρ = 1.6 kgf/cm^3) and a flow speed of 3 m/s the energy expended in vertical movement of a lump will be (if the speed of the lump is half the speed v_{s1})

$$P = \frac{mv^2}{2} = \frac{3 \cdot 1.5}{2} = 3.35 \text{ J} \qquad (8.5)$$

Usually enamel coatings have an impact strength of about 0.5 J, sitall and metal ceramic coatings have impact strengths of up to 10 J.

CHAPTER 9.   TECHNICAL AND ECONOMIC INDICES OF HYDRAULIC MINING
             AND HYDROTRANSPORT

9.1. Structure of Cost of Hydrotransport

At the present time, hydraulic transport can perform many tasks
involved in the movement of bulk materials over various distances;
numerous devices have been created for this purpose and methods of
their design have been developed. The task is to select from among
the many engineering versions, one which is characterized by minimal
cost, i.e., the task of finding the optimal parameters corresponding
to the minimum capital investment for construction and operating cost
of the hydrotransport system must be performed.

The optimal parameters can be defined by comparing versions, cal-
culating the cost figures for the hydrotransport installation given
various initial data -- consistency of the slurry, speed of its movement, various sizes of solid particles, etc.

Some factors, for example, the hydrotransport distance and throughput of the installation, cannot be changed or optimized. The most
important parameters which can be varied are the consistency of the
slurry and the speed of its movement. The particle size distribution
and maximum diameter of particles of the material transported can
also be changed in many cases. Determination of the most important
competing parameters which define the minimum of the adjusted costs
is an important task.

Proper economic evaluation of hydrotransport requires that the
analysis objectively consider the economic regularities and technical
capabilities for its utilization.

The structure of costs of hydrotransport is determined by the
features of the technological system and the devices used for preparation of the slurry, hydrotransport, reception and utilization of
the fuel.

The main expenditures which determine the cost of hydrotransport

221

of coal in hydraulic mines can be determined, as a result of planning studies and accumulated experience. We present herewith the
plan data accumulated by VNIIGidrougl' (table 9.1) for determination
of the economic effectiveness of the transportation of coal at the
Velovskaya regional electric power plant (Kuznets basin).

Figure 9.1 shows the cost structure of hydrotransport according
to foreign (US) data. The cost of dewatering and drying of the coal
has a decisive influence on the cost structure of hydrotransport
installations. The contradictory figures relating to the economics
of hydrotransport can be explained to a great extent by improper determination of these most important costs [26]. According to IGI, drying
and dewatering account for over 50% of all hydrotransport costs. If
the cost of preliminary dewatering of coal is taken as 100%, the
cost of dewatering in centrifuges is over 200%, drying costs exceed
500%.

Table 9.1

Cost items	Cost, rubles per ton	
	Hydrotransport	Rail transport
Dewatering:		
in the mine (loading)	–	0.757
at the power plant	0.534	–
Thickening of slurry	0.340	–
Hydrotransport	0.468	–
Nonproductive costs	0.094	0.094
Transport from mine to railroad station	–	0.196
Rail tariff for 30 km transport	–	0.540
Transport from railroad station to power plant	–	0.110
Unloading coal at power plant and transport to crushing section	–	0.320

222

The most complex and cumbersome process is that of dewatering
and trapping of fine slime (particle diameter less than 60 μ),
which is carried away with the centrifugate. Dewatering of 0 to 6 mm
coal is performed mechanically or in centrifuges. Coal less than
0.5 mm in diameter is dewatered on filter presses, then dried by
heat to the moisture content required by the power plant. The capital
investment in drying represents up to 40% or more of the total cost
of the dewatering system at a power plant; the operating costs may
be as great as 0.50 rubles per ton.

If heat drying could be eliminated, it would significantly improve
the cost figures of hydrotransport installations. Analysis of three
versions of heat drying, in drum dryers, in fluidized bed dryers and
in cyclone-type dryers, has established that approximately 3% of
the initial coal must be expended to generate the heat for drying,
regardless of the method or equipment used.

It must be considered that as of today only heat drying can provide
low coal moisture content (less than 10-12%). For coal hydrotransport
systems from hydraulic mines to large consumers, the cost of dewatering
and drying of the coal is not included in the cost of hydrotransport,
since the coal becomes wet during hydraulic mining.

The cost of dewatering and drying is great for all hydrotransport
installations; however, it is an additional cost for mines with dry
mining technology, since the wetting and drying of the coal in the
hydrotransport installation is not an inherent part of the mining
technology. For this reason, these installations may be undesirable.

Coal mined by a dry method may be transported by hydrotransport to
a beneficiation plant which uses the wet beneficiation process. The
effectiveness in this case may be high, if additional costs are not
incurred in reconstructing the beneficiation plant. Some such plants
are not suitable for receiving large quantities of slurry instead of
dry coal. If a consumer is receiving coal from a beneficiation plant
which uses the wet beneficiation process, the dewatering plant can be
constructed at the location of the consumer rather than at the beneficiation plant. Therefore, dewatering costs will not increase upon

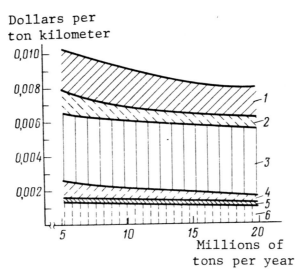

Figure 9.1. Cost structure of hydrotransport (according to US data): 1--capital investment; 2--cost of electric power; 3--crushing; 4--operation and maintenance; 5--amortization of electric equipment; 6--amortization of mechanical equipment.

transition to hydrotransport.

The capital investment for dewatering and drying depend little on the transportation distance (when finely dispersed slurry are transported) and the amount of crushing of the coal has very little influence on the cost.

The most important factor determining the required capital investment for dewatering and drying is the throughput of the hydrotransport installation. Dewatering and drying of coal are very expensive operations requiring significant capital investment and operating expenditures.

The effectiveness of hydrotransport is greatly influenced by the cost of receiving systems. Hydrotransport of coal over long distances is economically favorable where suppliers and consumers have great throughput capacity. A power generation unit with a capacity of 800 mW requires at least 2.5 million tons of natural coal per year. Usually a regional electric power plant has several such units. Therefore, a large regional power plant requires a coal pipeline with a capacity of at least 4-5 million tons of coal per year.

When large mines with the required output and large consumers are both present, individual receiving systems can be used, eliminating the cost for supplementary transport and allowing the use of emergency storage areas at the mine.

If no large suppliers are available, group receiving systems can be used. Railroad transport is used to deliver the coal to the systems.

The capital investment for the construction of receiving systems and dewatering and drying systems is great, up to 50-70% of the total capital investment, according to plan data.

The specific operating costs depend on the technological system used to receive the coal. With a throughput of 5.3 million tons per year and transportation of coal from the mines by rail, the specific share of these costs is as great as 40-56%. Hydraulic mines do not require receiving systems on the surface if the coal is hydrotransported directly to the power plant.

Hydrotransport is most effective when it is basically a continuation of a technological process, for example, when coal from a mine is transported directly to a large consumer. The use of hydrotransport as a connecting link between mine and consumer is most desirable.

The main factor determining the level of capital investment in hydrotransport is the quantity transported; the hydrotransport distance is less important. Calculations have shown that as throughput increases by a factor of 3.5, specific capital investments increase only by a factor of 1.5.

The economic results of hydrotransport of a given quantity of material over a given distance are greatly influenced by the particle size of the coal transported.

The influence of coal particle size on the economy of hydrotransport can be seen from table 9.2.

It is generally not desirable to transport large lump coal over long distances, since large particle sizes require high slurry movement speeds, resulting in increased consumption of power. Low slurry consistencies and the need to transfer ballast water mean that the capital investment increases and hydrotransport becomes undesirable.

225

Table 9.2

Coal particle size, mm	Pipeline diameter, mm	Capital invest-ment per ton of standard fuel, rubles	Operating cost per ton of standard fuel (with inhibitors), rubles
0-0.2	600	4.87	0.74
0-3	375	8.34	1.40

The cost of laying pipe may be as great as 50% of the total; therefore, increasing the service life of pipe is an important factor in increasing the effectiveness of hydrotransport. One means of increasing the service life of pipe is to increase the wall thickness, within tolerable limits.

A reduction in the cost of rotation and replacement of pipe decreases the amortization cost. The cost of pipe amortization can be decreased also by increasing its wear resistance (by using special materials and coatings) or by using pipe with variable wall thickness. The cost of hydrotransport installations should consider the cost of automation equipment, since complete automation of hydrotransport can significantly reduce personnel costs.

When the area of application of hydrotransport is properly selected, electric power costs are significant (40-43%).

The area of effective application of hydrotransport is basically where it is a necessary component of the technological process, i.e., where it is impossible to avoid it, for example, in hydraulic mines, where the process of hydraulic mining requires the use of hydrotransport. The process of wet beneficiation of coal and beneficiation plants allows hydrotransport to be considered a continuation of this technological process.

The effectiveness of long range hydrotransport depends more on throughput than on pipeline length; therefore, it should be used primarily for large consumers. The coal industry is developing in the direction of increasing the output of mines and thereby increasing

the productivity of labor. The tendency of the production of mines
to increase and the desire for effective utilization of hydrotransport
of coal over long distances are consistent in this case.

When preparatory and supplementary operations are included, the
effectiveness of long-range coal pipelines is determined not so much
by the cost of hydrotransport as by the cost of reception systems
and dewatering equipment. Hydrotransport systems also minimize preparatory and supplementary operations.

9.2. Approximate Method of Technical and Economic Design of Long-Range Hydrotransport

Economic analysis includes determination of the optimal speed of
flow of a slurry and coal pipe diameter for various hydrotransport
distances, throughputs and slurry density; optimal throughput of the
installation as a function of coal pipeline length; and optimal consistency of the slurry.

The optimality criterion for process parameters is the minimum overall (adjusted) cost of construction and operation of the long-range
hydrotransport installation.

Equations have been suggested for determination of the economic
indices of long-range hydrotransport of coal, based on experimental
data obtained for the movement of slurry in actual pipes and considering
the experience of planning of hydrotransport installations and transport
pipelines for oil and petroleum products both in our country and abroad.

In determining the optimal parameters, we assume that the operating
costs do not change from year to year, and that capital investments
are made at one time. The cost of hydrotransport of one ton of coal
for one kilometer

$$p_{T}'' = \frac{p''}{c_{B}}, \quad \text{rub} \cdot 10^3/\text{tkm} \tag{9.1}$$

where p" is the cost of hydrotransport of one tkm of slurry, $\text{rub} \cdot 10^3/\text{tkm}$
(all operating costs and capital investments are given in thousands of

rubles); c_B is concentration   of the slurry by weight.

The total   (adjusted) cost of hydrotransport of one tkm of slurry

$$p'' = \frac{k''}{T} + c''; \text{ rub.} \cdot 10^3/\text{tkm} \qquad (9.2)$$

where $k'' = \frac{k}{A_y L}$ represents the adjusted capital investment per tkm
for construction of the hydrotransport system; $c'' = \frac{c}{A_y L}$ represents
the adjusted operating cost per tkm; k is the capital investment for
construction of the hydrotransport system in each possible version;
c represents the operating cost in each version; A_y is the annual
throughput of the hydrotransport installation as slurry, $t \cdot 10^6/\text{hr}$, L is
the length of the pipeline, km; T is the standard amortization period,
in years.

The capital investment for construction of a hydrotransport system
can be divided into the capital investment in the initial operation
k_H and in subsequent operations k_p:

$$k = k_n + k_p; \qquad (9.3)$$

$$k_p = k_1 + k_g + n k_p , \qquad (9.4)$$

where k_1, k_g and k_p are the capital investment for the construction of
the line portion of the pipeline, the head pumping plant (HPP) and an
intermediate pumping plant (IPP); n is the number of pumping plants.

The capital investment in the initial operation is linearly dependent
on the throughput of the hydrotransport system

$$k_n = 4.18 \cdot 10^3 \rho_{s1} Q, \qquad (9.5)$$

where Q is the speed of the slurry, m/s.

The capital investment for construction of the line portion of the
pipeline depends on the throughput of the hydrotransport installation,
the speed of the slurry and the pressure. It can be determined from
the equation

$$k_1 = L \, [\, f(h) \, (D_y - 0.28) + 15.9],$$

(9.6)

where D_y is the standard inside diameter of the pipe, m; h is the head developed by the pumping plant, m water.

The function f(h) varies as follows:

h, m water . . .100, 150, 200, 250, 300, 350, 400, 450, 500, 550
f(h) 43.8 44.3 44.8 45.4 46.0 46.7 47.5 48.4 49.6 51.0

If all pumping plants develop the same head, then

$$h = \frac{H}{n + 1} \quad \text{m } H_2O$$

where H is the head necessary to overcome hydraulic losses over the length of the pipeline, m water.

The hydraulic losses in a pipe are determined by the equation

$$H = 1.05 \, i_{sl} \, L \cdot 10^3, \text{ m} \quad H_2O \, ,$$

(9.7)

where i_{sl} is the specific loss of head, determined by experiments performed in real pipelines using actual slurries.

The capital investment for the construction of the head and intermediate pumping plants depends on the throughput of the hydrotransport installation, capacity of emergency reservoirs, type and power capacity of pumps, and can be determined from the equations

$$k_g = f(m) + 953 \, \rho_{sl} Q + 260;$$

(9.8)

$$k_p = f(m) + 30.3 \, \rho_{sl} Q + 453.9,$$

(9.9)

229

where m is a parameter corresponding to the number of pumps installed at the HPP and IPP;

$$m = 0.034 \, \rho_{s1} Q \, h. \tag{9.10}$$

The function f(m) varies as follows:

m	2	4	6	8	10	12	14	16	18	20	22
f(m) . .	110	215	300	375	450	515	585	655	725	795	865

The operating cost can be expressed as the sum of the operating cost for the line portion of the pipeline, the head and intermediate pumping plants

$$E = E_1 + E_g + n E_p. \tag{9.11}$$

The operating costs for the initial operation are determined by the equation

$$E_H = 45 + 0.11 \, k_H + 5.35 \, (n + 1) \tag{9.12}$$

The operating cost for the line portion of the pipeline can be written as follows:

$$E_1 = 0.11 L + 0.07 \, k_1 \tag{9.13}$$

The operating cost for pumping plants is a function of the cost of electric power:

$$E_g = 25.84 + 0.0886 \, k_g + 0.025 \, f(m) + E_e + f(E_e); \tag{9.14}$$

$$E_p = 22.33 + 0.0886 \, k_p + 0.025 \, f(m) + E_e + f(E_e). \tag{9.15}$$

230

The annual cost of electric power is determined by the equation

$$E_e = 3.16 \cdot 10^{-5} A_y \frac{L}{n+1}. \tag{9.16}$$

The function $f(E_e)$ varies as follows:

E_e	50	100	150	200	250	300	350	400	450
$f(E_e)$	1,6	1,9	2,7	3,0	3,7	4,5	5,4	6,4	7,5

Calculations for many versions of coal pipelines have been undertaken for hydrotransport installations under the following conditions:

$A_y \cdot 10^6$, т/год. . . 1-2-3-4-5-6-7-8-9-10-11-12-13-14-15-16-18

L , км. 50-100-150-200-300-400-450-500-550-600

v , m/s 0,5-0,7-0,9-1,1-1,3-1,5-1,7-1,9-2,1-2,3-2,5

c 0,44-0,50

As an example, table 9.3 shows some values of the cost of transport of water-coal slurries (kopeks/tkm) with a slurry density of 1200 kg/m^3.

Table 9.3

Pipe diameter, mm	Throughput as slurry, $t \cdot 10^6$/yr			
	2	4	8	16
50	1,17-1,45	0,73-0,77	0,44-0,51	0,34-0,34
150	1,05-1,35	0,55-0,66	0,33-0,37	0,24-0,23
250	1,03-1,26	0,52-0,64	0,31-0,37	0,22-0,21
350	1,02-1,26	0,51-0,63	0,30-0,33	0,21-0,20
450	0,98-1,27	0,51-0,63	0,29-0,32	0,20-0,20
550	0,98-1,25	0,50-0,63	0,29-0,31	0,20-0,19
660	0,98-1,25	0,50-0,52	0,29-0,31	0,20-0,19

231

9.3. Influence of Technological Parameters on Cost of Hydrotransport

The adjusted cost of hydrotransport depends to a great extent on the throughput of the installation (figure 9.2). The greater the throughput of the installation, the lower the cost. Installations with a throughput (as slurry) of less than 4-5 $t \cdot 10^6$/yr are particularly expensive. The data obtained agree with the experience of operation of oil pipelines, where cost figures decrease with increasing pipe diameter.

As concerns hydrotransport distance, its influence on the adjusted cost is seen particularly clearly with transportation distances of 100-150 km (figure 9.3). Thus, as the transportation distance changes from 50-600 km, the cost of hydrotransport drops by a factor of 1.3-1.5; as the throughput increases from 1 to 18 million tons per year of slurry, the cost of hydrotransport drops by a factor of 6. It is obvious from this that hydrotransport is primarily suitable for high throughput installations. As concerns transportation distance, it may vary but should be over 100-150 km.

Calculations have been performed to determine the influence of the conditions of movement of the slurry (speed), specific head loss and consistency on the economics of long-distance hydrotransport. The calculations were performed by the method outlined above. The studies were undertaken for the following conditions: throughput of hydrotransport installation as slurry 8 million tons per year; transportation distance 430 km; slurry concentration by weight 0.34, 0.40, 0.44 and 0.50. The adjusted costs as a function of speed of highly concentrated slurry are presented in figure 9.4. The minimum of adjusted cost as a function of hydrotransport speed may not be very clear for high concentration water-coal slurries. This is because a water-coal slurry takes on the properties of an anomalous fluid and has significant initial shear stress. As concentration of the slurry decreases, the fluid loses the properties of an anomalous fluid and acts more

232

like a Newtonian fluid.

At higher speeds, the structure of the slurry breaks down. The specific head loss is increased slightly, for example, only by 10% as speed changes from 0.6 to 1.1 m/s (0.785-0.876%).

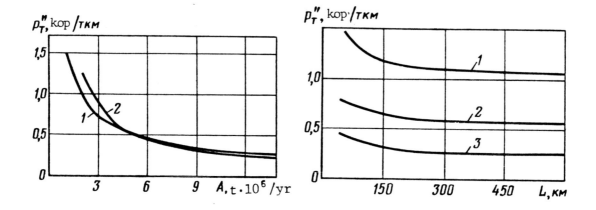

Figure 9.2. Influence of through-put of hydrotransport installation on adjusted cost: 1--c_p = 0.44, L = 100 km, v = 1.1 m/s; 2--c_p = 0.50, L = 100 km, v = 1.1 m/s

Figure 9.3. Influence of hydro-transport distance on adjusted cost: 1,2,3--throughputs of 1,2, and 5 $t \cdot 10^6$/hr

The optimal speeds for hydrotransport of coal at low slurry concentration are similar to the optimal speeds for pipeline transport of oil and petroleum products. The higher the concentration of slurry, the less clear is the minimum of cost for hydrotransport; in this case, preference should be given to higher speeds, since they decrease the cost of pipe and increase the reliability of operation of the hydro-transport system. Operating experience indicates that it is more reliable to have higher speeds, also considering the possibility of changes in the particle size distribution of the slurry.

233

It has been found that the optimal speed for transport of a water-coal slurry with a throughput of over 6 million tons per year is 1-1.5 m/s; when the throughput of the installation is less, the optimal speed is 0.9-1.3 m/s.

Reliable hydrotransport of water-coal slurries can be achieved with concentration by weight of 0.5 or even higher. However, this concentration cannot be considered economically most desirable; it is more favorable to transport slurries at concentrations up to 0.44-0.45 (figure 9.5).

Investigation of the influence of specific head losses on the cost of hydrotransport has demonstrated that the increase in adjusted cost is not directly proportional to the increase in specific head loss (figure 9.6). This is because the cost of hydrotransport is influenced not only by the cost of electric power, which depends directly on specific head losses, but also by a number of other factors. The figures presented indicate that the economic effectiveness of hydrotransport depends to a great extent on proper selection of installations, their throughput, as well as the technology of use of the coal.

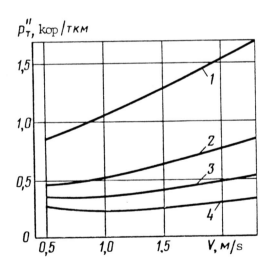

Figure 9.4. Influence of speed of hydrotransport on adjusted cost: 1-4--throughputs of 1,2,3 and 5 million tons per year; L = 400 km; c_p = 0.5

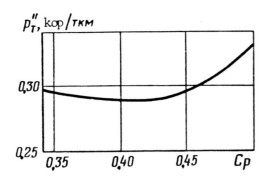

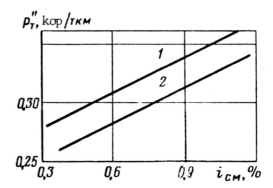

Figure 9.5. Influence of slurry
consistency on adjusted cost

Figure 9.6. Influence of specific
head loss on adjusted cost: 1--c_p =
0.5; A = 8 t·10^6/yr; L = 430 km;
2 - c_p = 0.445; A = 8 10^6/yr; L =
430 km

In selecting hydrotransport installations it is quite important to
consider the cost of preparatory and supplementary operations, and when
large-lump abrasive materials are transported the costs resulting from
wear of equipment and pipes, as well as the additional cost related to
the reduction in value of the material due to attrition during transportation. These costs may be great and may increase the cost of the
process of hydrotransport itself.

CHAPTER 10. PROSPECTS FOR THE DEVELOPMENT OF HYDRAULIC MINING AND HYDROTRANSPORT

10.1. Area and Volume of Application of Hydraulic Mining

Recommendations have now been produced for the development of a hydraulic mining and hydrotransport technology for coal. The prospects for development of hydraulic mining have been determined primarily for power coal, to be burned in powder form at electric power plants.

The hydraulic technology has been extensively tested in industry, so that the main technological systems can now be recommended:

gravity transport through the shaft mine from the mining face to a central hydraulic hoist, with openings up to 4 km in length; where the openings are longer, pressurized hydrotransport should be used;

hydraulic hoisting with coal pumps, airlifts and feeders;

when it is necessary to extract the larger lumps, these large lumps can be transported by conveyors or skip hoists, the smaller particles being transported by coal pumps, feeders or airlifts;

mining of coal with series-produced monitors, remotely controlled, water pressures up to 12 MPa, with subsequent introduction of hydraulic breaking at water pressures up to 16 MPa;

mining of coal with mass-produced mechanical-hydraulic mining machines, remotely controlled, in short mining faces;

driving of openings with mass-produced mechanical-hydraulic combines, remotely controlled;

mass-produced equipment for wet beneficiation plants with an increase in the length of the flotation system by up to 6%, should be used for beneficiation, dewatering and drying of coal and clarification of recycled water.

The effectiveness of operation of the beneficiation plant can be increased by increasing the consistency of the slurry transport to 1:5, which has already been achieved at the best shaft mines in the Kuznets basin.

236

Long-range hydrotransport from the mine to large consumers is desirable, using a hydraulic mine -- coal pipeline -- consumer system.

Long-range hydrotransport of coal allows: a decrease in the cost of transportation of coal in comparison to rail transport; elimination of labor consuming rock loading operations and pollution of the environment; assurance of high reliability of the hydrotransport system in terms of throughput.

Studies by UkrNIIGidrougl', VNIIGidrougl', IGI and other institutes, involving many years of research, have resulted in the creation of the scientific principles of long-range hydrotransport of coal. It has been established that this type of transport is particularly effective when it is a component part of a unified technological process of hydraulic mining with high capacity of the system. Suggestions have been prepared for the planning of large hydrotransport installations to feed coal from the hydraulic mines of the Kuznets basin to the Perm' regional electric power plant. A fuel and power system including the mines of the Dobropol'skiy rayon and a coal pipeline to the Chigirinskaya regional electric power plant is now in the planning stage. Suggestions for the hydrotransport of coal from the Ekibastuz (Maykyubinsk) mine to the Balkhashsk regional electric power plant are now being developed. The suggestions call for transport of coal and hydrotransport of slurry with a heat of combustion of 2800-4000 kcal/kg.

The Kuznets basin is to be the site of an extensive program of development of hydraulic technology. Progressive technological systems which have been developed are now to be put into practice. Plans include modernization and introduction of new equipment which will increase the productivity of labor of a worker in the Kuznets basin. Extensive work is under way on the creation and assimilation of new machines and equipment: the 12GD-2 monitor, MGPP tunneling machine, DKV1 crusher-classifier installation and other devices.

Over the past ten years the volume of hydraulic mining and hydrotransport in the Donets basin has increased by a factor of 2.2. Hydraulic mines have reached their full planned capacity, and the productivity of labor has reached 60.9 tons per month. Hydraulic mining and hydrotransport systems have been equipped with modern equipment, the reliability of which is two to three times greater.

The experience which has been accumulated indicates that hydrotransport is the basis for hydraulic technology. The advantages of this technology for underground mining of coal under difficult geologic conditions has been proven.

The redesign of "Pioner," "Krasnoarmeyskaya" ("Dobropol'yeugol'") and the 50th anniversary of the USSR mine ("Krasnodonugol'") are planned, with an increase in the output capacity of each of these mines.

UkrNIIGidrougl' is working to replace short-run railroad transportation of coal with pipeline hydrotransport from the mines of "Makeyevugol'," "Krasnoarmeyskugol'" and "Donetskugol'" unions to the Kolosnikov and Kurakhov central beneficiation plants of "Donetskugleobogashcheniye" union.

Scientific research work on the creation of equipment and processes for unattended mining of coal is to be continued, with particular attention to processes suitable for mines in which there is danger of explosion, improvement of the technology and equipment for hydrotransport of coal to large consumers, improvement of processes for dewatering of coal and clarification of process water in the mine and on the surface, and many other areas.

In 1979, planning was begun on a Kuznets basin-Novosibirsk coal pipeline about 250 km in length.

10.2. Basic Trends in Development

Large, highly productive mechanized and completely automated hydraulic shaft mines, directly connected by large pipelines to coal-fired electric power plants, coal-tar chemical plants or other large consumers can be considered mines of the future.

It is desirable to construct these hydraulic mines as a component part of a fuel-energy or fuel-metallurgical system. This allows the greatest economic effect to be achieved.

The primary trend in the development of hydraulic mining and hydrotransport of coal must be considered the creation of fuel-energy and fuel-metallurgical systems consisting of a mine, coal section, beneficiation plant, coal pipeline and a large consumer.

Hydraulic transport, one of the main links in these systems, is most suitable for transport of large quantities (several millions of tons per year) of finely ground coal as finely dispersed slurry, similar in its properties to liquid fuel.

The hydrotransport of coal within the USSR can be implemented as follows:

transport of large volumes of coal over long distances. Of particular value is main line hydrotransport of brown coal from the Kansk-Achinsk basin to the center of the nation. Hydrotransport cannot be replaced with any other type of transportation for this purpose;

transportation of coal from permanent suppliers to permanent consumers, for example, from mines to beneficiation plants and from beneficiation plants to coal-tar chemistry and metallurgical plants. Elimination of the "short legs" can greatly reduce the losses of coal during transportation as well as the cost of loading and unloading operations;

creation of fuel-energy and fuel-metallurgical systems with coal or coking products fed to large consumers. The reduction in transport costs and refinement of technological processes result in a decrease in the cost of electric power at coal-fired power plants or the cost of metal at metallurgical plants;

improvement of hydrotransport within hydraulic mines and use of this method for the transportation of materials for hydraulic stowing.

The USSR coal industry ministry is working on the planning of long distance coal hydrotransport systems to connect coal mining enterprises to large consumers in order to reduce the rail transport of coal. The capabilities of pipeline transport are being considered, in that it provides continuity and regularity of delivery of large quantities of

239

material, eliminating cumbersome loading and unloading operations by the
supplier and consumer, reducing the danger of environmental pollution and
eliminating the loss of coal during transport while allowing the creation
of completely automated high capacity transportation systems at lower
cost than traditional methods of transportation.

The necessary scientific-technical and economic data is being accumulated for planning and construction of fuel-energy systems in the next
few years, in which coal from large suppliers will be transported over
distances of 200-300 km or more to large consumers. The capacity of the
existing hydrotransport system from "Yubileynaya" hydraulic mine to the
western Siberian metallurgical plant is to be more than doubled.

Super long-range hydrotransport systems are being developed to
supply coal from the Kansk-Achinsk basin to various sections of the country
(hydrotransport distance 2500-4000 km).

10.3. Conditions of Effective Application of Long-Range Hydrotransport

The area of effective utilization of hydrotransport of coal is
limited by technical and economic factors.

The limiting technical factors are: the need to preserve the grading
of the coal; the need to prevent contact between certain materials which
are transported and the carrier fluid. Hydrotransport therefore cannot
be recommended for materials whose useful properties deteriorate as they
are carried by fluids through pipes or flumes.

Limiting economic factors include:

the process of hydrotransport of the material requires that it be
wet, then dewatered and dried by the consumer, the cost of dewatering and
drying possibly exceeding the cost of hydrotransport itself;

highly abrasive materials may cause excessive cost for replacement
of coal pipes and pumps;

when large lump materials are transported, the power consumption of
the process may be excessive;

when small quantities of coal are transported, the costs of maintaining receiving sections become excessive.

Hydrotransport in hydraulic mines involve transportation over relatively short distances (3 to 5 km) of slurries of relatively low concentration by weight (not over 5-10%) moving at 2-4 m/s, with particle sizes of up to 70-100 mm, requiring a power consumption of 2 to 3 kW·hr/tkm.

The effectiveness of long-range hydrotransport of coal depends to a great extent on the technological parameters of the process and the quality of the equipment used. Some of the process parameters, for example, the throughput of the installation and the transportation distance, cannot be arbitrarily selected, but are rather determined by the task at hand. Others, such as the particle size of the coal transported, speed of movement and slurry consistency, must be selected following technical and economic calculations.

The effectiveness of long-range hydrotransport is influenced by the durability of coal pipelines and equipment, as well as the degree of automation.

One of the most important parameters influencing capital and operating costs is the throughput of hydrotransport installations. The minimum throughput at which long-range hydrotransport is effective can be determined in the planning of specific projects. It can be assumed that the throughput should be at least 5-10 million tons per year, at which point it is expedient to operate slurry preparation, crushing and dewatering installations as well as the main hydrotransport line itself.

If the throughput is quite small, the effectiveness of hydrotransport drops rapidly, since it becomes necessary to construct reserve containers to allow operation of the installation with interruptions or to support continuous operation through small diameter pipelines; then other processes are economically expedient.

The area of application of the process of hydrotransport is determined by the mining conditions, or the conditions of further processing of the particulate material.

When coal deposits are hydraulically mined, hydrotransport is obligatory, and cannot be replaced by any other method of delivery of the slurry to the consumer.

Hydrotransport is also irreplaceable if the technology of processing of the material requires that it be wet, for example when coal is transported to a wet beneficiation plant.

Main line hydraulic transport systems are characterized by great length (a coal pipeline 440 km in length is now in operation, and plans for hydrotransport systems up to 4000 km are under development), high concentration of the slurry (up to 45-50% by weight), low movement speed (1.1-1.8 m/s), small particle size of the material transported (up to 2-3 mm) and low power consumption (not over 40-60 W·hr/tkm).

Hydrotransport is particularly favorable when it is possible to reduce the cost of loading and unloading operation, where the quantity of material to be transported is large and water is present in sufficient quantities at the loading point, when hydrotransport precedes the technological operations of wet beneficiation of coal or follows these operations.

Hydrotransport has great advantages over other means of energy transport, including rail transport or electric power transmission lines, from the standpoint of protecting the environment from possible pollution. Pipelines are laid beneath the surface, are completely covered, and the surface above them can be used, since the pipe does not need to be changed for long periods of time.

Hydraulic transport through pipes is one of the most reliable types of transportation, since it is not subject to various unfavorable influences of the weather. It can be completely mechanized and automated, and requires minimum operating personnel. Its reliability is greater than that of other types of transport (for example, the "Black Mesa" coal pipeline was down a total of 33 hours in a year, its reliability is 99%).

Pipeline transport of finely ground coal has high reliability, since water-coal slurries of this type are liquid fuels, capable of transportation at very low speeds with low power consumption and very little wear of pipes and equipment.

As operating experience has shown, main line hydrotransport is very significant for the transportation of large quantities of fuel; it is then possible to decrease specific capital investments, decrease transportation costs and increase the productivity of labor. In some cases hydrotransport of coal is cheaper than transmission of electric power.

Hydraulic transport is effective if it is a continuation of the process of hydraulic mining of the coal; the tendency toward construction of large hydraulic mines matches the tendency toward increasing effectiveness and utilization of hydrotransport.

Systems in which hydrotransport is a continuation of the process of hydraulic mining and in which the preparation of the slurry for hydrotransport is not a special process, but rather a part of the general technological process, are most promising for hydrotransport. Wherever the technological process produces a slurry, the use of hydrotransport is the only proper answer.

Hydrotransport can also be used where there is no space for the construction of railroad lines, where pollution of the environment must be prevented, and where safety requires its use.

The overall effectiveness of hydrotransport, including dewatering and drying installations, required by the technology of hydrotransport, must be determined not only considering the costs of hydrotransport, but also the cost for dewatering and drying of the coal.

If the coal arrives at a power plant as a water-coal slurry from a wet-process beneficiation plant, the process of drying is shifted from the beneficiation plant to the power plant, and the costs of dewatering and drying are not additional costs of hydrotransport. Therefore, the effectiveness of hydrotransport is quite high. When additional costs are incurred for dewatering and drying of the coal, the effectiveness of hydrotransport is greatly reduced, since these costs may vary from 33% to as much as 63% of the total operating costs.

In properly selected hydrotransport systems, the process parameters must be properly selected -- particle size distribution of coal, maximum coal particle diameter, speed of movement and density of the slurry, efficiency of pumps, i.e., the system must be made to operate at the minimum total cost.

As studies have shown, hydrotransport over long distances is best performed with water-coal slurries containing over 25-30% coal particles larger than 0.063 mm with a maximum particle diameter of 1.5-2 mm.

These coarsely dispersed slurries require, in comparison to finely dispersed slurries (containing only 5-10% particles larger than 0.063 mm), less power for grinding of coal; they move at low speeds with low specific head losses; and they assure normal operating conditions for the hydrotransport system.

When the system is shut down, coarsely dispersed slurries separate into solid and liquid phases. The speed of separation depends on the dispersion of the solid medium, the consistency of the slurry, its temperature and other factors. The sediment formed upon layer separation is loose and mobile, and has some of the properties of a solid, such as a yield point. The sediment, due to its weak structure, is easily restarted in motion, assuring normal restart of the hydrotransport system even after a long shutdown.

Coarsely dispersed slurries of low consistency, moving through large diameter pipes, act very much like homogeneous fluids: there is no structure to the fluid or if it is present, it is easily disrupted by mechanical force, even that resulting from the slow movement of the system.

Coal is transported over long distances to coal-fired electric power plants, beneficiation plants and coal-tar chemical plants. Coal intended for power production which is not subjected to beneficiation is suitable for all known methods of hydromechanization. Power coal which is beneficiated must be protected from excessive attrition, since the yield of the 0-1 mm fraction is greater following hydromechanization than at mine beneficiation plants using dry processes, and the cost of beneficiation of the smaller particles is significantly higher than the cost of treating larger particles. It must be recalled that hydraulic mining and hydrotransport of coal as a slurry to the beneficiation plant results in production of coal which is not as greatly ground in beneficiation as coal from dry process mines. This is because the water at hydraulic mines breaks down the particles of coal and forms smaller particles, whereas in dry mines all of this process occurs at the beneficiation plant.

Let us look at the requirements for hydrotransport of coal to coal-tar chemical plants. Run-of-mine coal is beneficiated at a beneficiation plant and a mixture is prepared which is then hydraulically transported to the coal-tar chemical plant, where it is dewatered, dried, then coked.

UkrNIIGidrougol' Institute has studied the grinding of coke concentrate from the Chumakovskaya central beneficiation plant during long distance hydrotransport. The studies were performed using a 0-6 mm mixture with the same screen composition as is presently used. It was found that with distances of 300-600 km, the content of the 0-3 mm fraction increased from 89.5% in the initial product to 95.8% of the final product. The content of 0-1 mm particles increased from 52.5% of the initial product to 73.8% of the final product.

A number of studies have established that high quality coke can be produced from a charge containing over 95% of the 0-3 mm fraction. Thus, hydraulic transportation of coal to coal-tar chemical plants must be considered as a usable, and in many cases an effective, form of transport.

10.4. Examples of Possible Application of Long-Range Hydrotransport

The experience which has been accumulated in planning and operation of systems allows the development of installations for long-range hydrotransport to be developed today for various conditions.

Large super long distance coal pipelines could play a significant role in the development of the economy of the USSR, by allowing at least 250 million tons of cheap coal to be transferred from the Kansk-Achinsk region to areas of utilization within the foreseeable future. The transportation of such tremendous quantities of energy is such a huge problem that it cannot at present be performed sufficiently effectively by any existing form of transport.

Kansk-Achinsk brown coal has certain properties making its transportation by rail quite difficult. This is a spontaneously burning coal with high natural moisture content. It is difficult or impossible to transport

brown Kansk-Achinsk coal in ordinary railroad cars, and quite expensive
to transport it in special sealed cars filled with inert gases.

An ordinary railroad line could transport up to 100 million tons
of natural fuel per year; therefore, a special coal transportation super
main line would be required to transport 250 million tons of natural fuel
per year. Railroad transport involves the loss of fuel and poisoning of
the atmosphere.

Furthermore, the construction of large electric power plants in the
central region of the nation cannot be tolerated unless problems of environmental protection and utilization of the ash and slag generated
can be solved.

The methods of transportation which we have studied have their advantages and disadvantages and do not meet all the requirements placed upon
them. It is therefore desirable to seek out a combination of various types
of energy carriers and methods for their transportation considering their
specific features. It has been suggested that electric power be transported to the center of the nation. This would allow us to make use of
its properties of easy divisibility; the problem of utilization of the ash
and slag material and of pollution of the environment in densely populated
areas would then not be direct problems for the consumers.

Electric power can be generated at coal-fired electric power plants
1500-2500 km from the center of the country. This would simplify the
problems of the electric power industry related to super long-range electric
power transmission. There is no need to use super high voltages, which
would decrease capital investment and increase the effectiveness of transportation.

Coal fired electric power plants, it has been suggested, should
be located at some distance from the center of the nation, and should
not be concentrated in a single area. Locations should be selected considering the possibility of creating enterprises to utilize the ash and
slag, the problem of atmospheric pollution, etc.

The figures relating to hydrotransport of 50 million tons of
Kansk-Achinsk coal to the center of the nation each year are presented
in table 10.1.

Table 10.1

Items	Length of coal pipeline, km		
	2000	3000	4000
Annual volume of coal, $t \cdot 10^6$	50	50	50
Annual volume of slurry, $t \cdot 10^6$	84	84	84
Throughput of hydrotransport system, m^3/hr	9252	9252	9252
Standard diameter of pipe, m	1.4	1.4	1.0
Specific head losses, %	0.42	0.42	0.42
Hydraulic losses in pipeline, m water	8820	13230	17640
Head per pumping plant, m water	640	640	640
Spacing between pumping plants, km	142.9	142.9	142.9
Number of pumping plants	14	21	28

Economic calculations indicate the advantages of main line hydraulic pipeline transport over other types of transportation to deliver the inexpensive Kansk–Achinsk basin coal to the central European portion of the USSR.

The super long-range Kansk–Achinsk coal pipeline can be planned using the experience gained in planning the 1666 km pipeline in the USA, where coal from deposits in Wyoming is to be transported hydraulically to a coal-fired electric power plant in Arkansas. The coal is prepared for hydrotransport in special crusher complexes, beneficiated, transported by six piston pumps through a 0.97 m diameter pipeline to a terminal point, where it is dewatered, dried and fed to the steam boilers.

An interesting coal pipeline project 420–450 km in length for the transportation of 4.2 million tons of coal from the southern Donets basin mines to the Black Sea region has also been developed.

Experience has been gained in the creation of coal pipelines of this length abroad; the projects which have been developed in the Soviet Union are listed in table 10.2.

Table 10.2

Items	Southern Donets Basin- Black Sea coal pipeline suggested by UkrNIIGidrougol')	Dnepr regional power plant pipeline (planned by "Teploelekt- roproyekt")	Existing "Black Mesa" pipeline in USA
Throughput, $t \cdot 10^6$/yr	4.2	4.3	4.8 (5.0)
Length, km	420-450	436	440
Mean diameter of solid particles, mm	1.3	0.1	1.3
Consistency of slurry, %	50	50	50
Pump type		piston	

The mass of pipe required to construct the southern Donets Basin-Black Sea pipeline is $45.4-48.6 \cdot 10^3$t. The approximate cost of the coal pipeline alone, including laying and installation of the pipe, is 23.8-25.5 million rubles. The total cost of the entire system will include the cost of receiving equipment at the Black Sea port.

One possible area of application of hydrotransport is the transportation of coal to coal carrier ships for further transport by sea. A ship loading system called "Marconaflo" is presently in use for the transportation of bulk cargo, including coal.

The transportation of coal by sea and the length of such voyages are increasing in our country from year to year. The increasing volume of the transportation has resulted in an increase in the tonnage of ships used to transport bulk cargo and the mean length of bulk cargo voyages by 270 km each year, i.e., by more than 2.5% per year.

As the tonnage of bulk cargo carriers increases, the significance of the problem of waiting time of ships grows. The mean waiting time of a ship in a coal port has reached eight days. The cost of loading of coal in ports is therefore increasing, in spite of the prevalence of highly productive loading equipment. The cost of handling of one ton in American ports has risen from 1.81 to 3.02 dollars (between 1950 and 1968).

Hydrotransport of coal in combination with marine transport by ships is quite promising. The essence of the method of marine transport of coal as slurry is as follows. Coal from a large supplier is transported by the hydraulic method as a slurry directly to the port and pumped into the hold of the ship, where the particles of coal settle to the bottom, and the excess water is then pumped out. After the ship arrives at its destination, the coal slurry is adjusted to the consistency necessary for transportation and is then transported through pipes to a storage area.

The coal particles to be transported should be not over 13 mm in diameter, the throughput capacity of the system quite high (at least 6 million tons of coal per year). The coal can be supplied by hydraulic mines or dry mines; in the latter case, equipment for preparation of slurries is required. A coal slurry with a concentration of up to 45-50% is transported by underwater pipeline to ships in the roads. As the coal separates and the water is drained, some of the fine particles may be lost. A method has been developed in Japan for purifying the water so that it can be dumped into the sea right in the port. This purification is achieved using coagulants, which greatly reduce the content of impurities. The method of marine transportation in ships allows a continuous process of transportation to be created, combining highly economical sea transport with pipeline transport. The great advantages of hydrotransport, such as no loss of cargo during transport and loading, simplicity of placement of cargo, high throughput and others can be supplemented by the great economic advantages of sea transport.

Achievement of highly economical slurry transport by sea requires study of the most efficient and least expensive methods to achieve rapid sedimentation of solid particles. The method of sea transport of coal as slurry may be used in our country, particularly when coal is exported, but also for transportation within the nation.

REFERENCES

1. Gidravlicheskaya dobycha uglya v Kuzbasse v IX pyatiletke (Hydraulic Mining of Coal in the Kuznets Basin During the Ninth Five Year Plan), Novokuznetsk, 1976 (VNIIGidrougol').

2. "Hydromechanization of Hydrotransport in Coal Mines," Trudy VNIIGidrouglya, No. 35, Novokuznetsk, 1975.

3. Krivchenko, A.A., Tsyapko, N.F., "Improvement of Hydraulic Breaking of Coal," Ugol', No. 9, 1964, p. 16

4. Spravochnik po obogashcheniyu ugley (Coal Beneficiation Handbook), edited by I.S. Blagov, A.M. Kotkin and N.A. Samilin, Moscow, Nedra Press, 1974

5. Volkovyskiy, Ye.G., Shuster, A.G., Ekonomiya topliva v kotel'nykh ustanovkakh (Fuel Economy in Boiler Installations), Moscow, Energiya Press, 1973

6. Andreev, S.Ye., Zverevich, V.V., Perov, V.A., Drobleniye, izmel'cheniye i grokhocheniye poleznykh iskopayemykh (Crushing, Grinding and Screening of Useful Minerals), Moscow, Nedra Press, 1966

7. Ofengenden, N.Ye., Izmel'cheniye uglya pri gidravlicheskom transporte (Attrition of Coal During Hydraulic Transport), Moscow, Gostoptekhizdat Press, 1962

8. Ofengenden, N.Ye., Svyatokaya, M.G., Andreeva, I.F., "Attrition of Coal During Hydraulic Mining and Hydrotransport," in Voprosy dobychi uglya gidravlicheskim sposobom (Problems of Hydraulic Mining of Coal), Moscow, Gosgortekhizdat Press, 1961

9. Ofengenden, N.Ye., Goshtovt, V.I., Grigoryuk, Ye.V., "Study of Hydrotransport of Water-Coal Suspensions in Large Diameter Pipelines," Gidravlicheskaya dobycha uglya (Hydraulic Mining of Coal), 1969, No. 12, p. 12

10. Geyer, V.G., Kostanda, V.S., "Universal Characteristics of the Hoist Pipe of an Airlift," Gidravlicheskaya dobycha uglya, Moscow, Nedra Press, 1965, No. 4, p. 30-34

11. "Operational Characteristics of an Airlift Hydraulic Hoist Operating With Slurry," V.G. Geyer, V.I. Gruva et al, Gidravlicheskaya dobycha uglya, 1966, No. 6, p. 37-40.

12. "Study of Operating Conditions of a Hydrotransport Installation With Sluice Feeders," B.M. Shkundin, V.M. Freydin et al., in Dal'niy truboprovodnyy gidrotransport sypuchikh materialov (Long Distance Pipeline Hydrotransport of Particulate Materials), Trudy AN GSSR, Tbilisi, 1974, p. 175-184

13. "Unsteady Movement of Two-Phase Media in Pipes With a Change in Pressure Drop With Time," Faizulayev, D.F., Umarov, A.I., et. al., Voprosy Mekhaniki, No. 8, 1970, Tashkent, FAN Press, UzSSR, p. 27-32

14. Dzhvarsheishvili, A.G., "Inertia Phenomena in the Suction Pipes of Piston Pumps," Soobshcheniya AN GSSR, Tbilisi, Vol. 55, No. 2, 1969, p. 353-355

15. Ofengenden, N.Ye. Collection No. 30, DonUGI, 1963, Moscow, Gosgortekhizdat Press, p. 404-417

16. Fyurer, G., "Freely Suspended Self-Supporting Pipes in Shafts," Gluckauf, 1967, No. 6, p. 19-28

17. Kamenev, V.V., "Stability of Vertical Pipes With Gland Compensators in Deep Mine Shafts," Trudy VNIIGidrouglya, No. 27, Novokuznetsk, 1973, p. 165-169

18. Dzhvarsheyishvili, A.G., Kirmelashvili, G.I., Nestatsionarnye rezhimy raboty sistem podayushchikh dvukhfaznyyu zhidkost' (Unsteady Operating Conditions in Systems Transporting Two-Phase Fluids), Tbilisi, Metsniyereba Press, 1965

19. Lunyakina, T.B., "Influence of Friction on the Amplitude of a Direct Hydraulic Shock," Sbornik trudov TBIIZhT, No. 31, Tbilisi, 1958, p. 71-89

20. Dzhvarsheishvili, A.G., Turabelidze, V.G., Tsamalashvili, T.Sh., "Hydraulic Shock in Two-Layer Pipelines," Trudy Bolgarskoy AN. Teoreticheskaya i prilozhna mekhanika, Sofiya, 1975, VI, No. 2, p. 77-81

21. Makharadze, L.I., Dzhvarsheishvili, A.G., Kirmelashvili, G.I., "A Device for Protection of a Pipeline and Hydrotransport System With Pumps From Hydraulic Shock," USSR Authors' Certificate No. 355313, Bulletin No. 31, 16, 10, 1972, p. 21

22. Serov, A.V., Diagnostika i upravleniye sostavleniyem sistem (Diagnosis and Control of the Condition of Systems), Moscow, Znaniye Press, 1974

23. Ekber, V.Ya., Kostovetskiy, S.P., Gontov, A.Ye., "Industrial Assimilation of New Technology, a Prerequisite for the Creation of Mines at the Next Technical and Economic Levels," Trudy VNIIGidrouglya, No. 36, Novokuznetsk, 1975, p. 125-131 (VNIIGidrougol')

24. Orfeev, Yu.V., "The Influence of the Operating Conditions of a Coal Pump on the Life of the Rotor," Trudy VNIIGidrouglya, No. 29, Novokuznetsk, 1975, p. 49-52 (VNIIGidrougol')

25. Slobodinskii, I.N., "Technical Details of the Manufacture of the Impellers of High-Head Pumps for Hydraulic Mines of Wear-Resistant Alloys," Trudy VNIIGidrouglya, 1974, p. 52-64 (VNIIGidrougol')

26. Ofengenden, N.Ye., Churkin, F.I., "Economic Effectiveness of Long-Range Hydrotransport of Coal," Ugol' Ukrainy, No. 11, 1965, p. 10-11

27. Ofengenden, N.Ye., "Determination of the Parameters of Long-Range Hydrotransport of Particulate Materials," Gidromekhanika, No. 25, Kiev Naukova dumka Press, 1973, p. 19-25

Western Office,
Mining Research Laboratories,
Canmet,
3303 - 33 St. N. W.,
Calgary, Alberta.
T2L 2A7